T0225030

China's Big Science Facilities

"Big science" facilities are major elements of science and technology infrastructure, and important symbols of China's scientific and technological development. This popular science book series presents the background, history and achievements of the Chinese Academy of Sciences in terms of constructing and operating big scientific facilities over the past few decades.

The series highlights the major scientific facilities constructed in China for pioneering research in science and technology, and uses straightforward language to describe the facilities, e.g. the fully superconducting Tokamak fusion test device (EAST), the National Protein Science Research Facility, Lanzhou Heavy Ion Accelerator, Five-hundred-meter Aperture Spherical Telescope (FAST), etc. It addresses the respective facilities' research fields, scientific backgrounds, technological achievements, and strategic and fundamental contributions to science, while also discussing how they will improve the development of the national economy. Supplementing the main text with a wealth of images and linked videos, the book offers extensive information for members of the general public who are interested in scientific facilities and related technologies.

More information about this series at http://www.springer.com/series/16530

Rendong Nan

Editor

The Sky Eye

Five-Hundred-Meter Aperture Spherical
Radio Telescope (FAST)

ZHEJIANG EDUCATION PUBLISHING HOUSE

Editor
Rendong Nan
National Astronomical Observatory of China
Chinese Academy of Sciences
Beijing, China

Translated by
Xiaobing Chen
Beijing Foreign Studies University
Beijing, China

Qiuju Huang
Beijing Foreign Studies University
Beijing, China

ISSN 2662-768X   ISSN 2662-7698 (electronic)
China's Big Science Facilities
ISBN 978-981-16-3826-8   ISBN 978-981-16-3824-4 (eBook)
https://doi.org/10.1007/978-981-16-3824-4

Jointly published with Zhejiang Education Publishing House
The printed edition is not for sale in China Mainland. Customers from China Mainland please order the print book from Zhejiang Education Publishing House.

Translation from the Chinese Simplified language edition: 观天巨眼——五百米口径球面射电望远镜 (FAST) by Rendong Nan et al., © Zhejiang Education Publishing Group 2018. Published by Zhejiang Education Publishing Group. All Rights Reserved.

This Springer imprint is published by the registered company Springer Nature Singapore Pte Ltd.
The registered company address is: 152 Beach Road, #21-01/04 Gateway East, Singapore 189721, Singapore

Editorial Board

Series Foreword

As a new round of technological revolution is bourgeoning, it will exert a direct impact on survival of a country whether or not it can gain insight on the future technological trends and grasp new opportunities from the revolution. In face of the major opportunities in the twenty-first century, China is intensively formulating the innovation-driven development strategy and building an innovation-based country in this critical era to achieve a moderately prosperous society in an all-round way.

Scientific and technological innovation and popularization remain two wings for innovation-driven development of a nation. In particular, popular science affects the awareness of the general public for science and technology as well as social and economic development. Scientific education is thus highly practical for implementing the innovation-driven strategy. Contemporary science pays more attention to public experience and engagement. The word "public" covers various social groups that exclude those in scientific research institutions and departments. The "public" also includes decision-makers and management personnel in government agencies and enterprises, media workers, entrepreneurs, science and technology adopters, etc. Barriers that impede the innovation-driven strategy will emerge if any group falls behind this new revolution; avoiding and removing the possible barriers will strategically improve the quality of human resources, enhance mass entrepreneurship and innovation and build a moderately prosperous society in an all-round way.

Science workers are primary creators of scientific knowledge who undertake the mission and responsibility for science popularization. As a national strategic power in science and technology, Chinese Academy of Sciences (CAS) has always attached equal importance to this mission in addition to scientific innovation and incorporated the mission into key measures of the "Pioneering Action" Plan. CAS enjoys rich and high-end technological resources, such as the high-caliber experts represented by CAS members, advanced research facilities and achievements represented by the Big Science Project and excellent scientific popularization base represented by the national scientific research and popularization base. With these resources in place, CAS implements the "High-level Scientific Resource Popularization" Plan to transform the resources into popular facilities, products and talents to benefit trillions of the public. Meanwhile, CAS launches the "Science and China" program, a scientific education plan, to mobilize more effectively the "popularized high-end scientific research resources" for scientific education targeted at the public and the integration of science and education.

Scientific education requires not only dissemination of scientific knowledge, approaches and spirit to enhance overall scientific literacy of the country, but also creation of scientific environment to enable scientific innovation to lead sustainable and sound social development. For this reason, CAS cooperates with Zhejiang Education Publishing House to launch the CAS Scientific and Cultural Project. This is a large-scale scientific and cultural communication project on the basis of CAS research findings and expert teams to improve the scientific and cultural quality of the Chinese citizen in an all-round manner and to serve for the national strategy of rejuvenation by advancing science and technology. On the basis of the target group, the project is categorized to two series, i.e., the Adolescent Scientific Education and the Public Scientific Awareness, respectively, for the adolescent and the general public.

The Adolescent Scientific Education series aims to create a series of publications that draw on latest scientific research findings and introduce the status quo of scientific development in China; to cultivate the adolescent's interest in science study; to educate them about basic scientific research approaches; and to inspire them to develop rational scientific way of thinking.

The Public Scientific Awareness series aims to educate the general public about basic scientific approaches and the social significance of science and encourage the public to engage in scientific affairs, thus the project will enhance the capacity the public of conscientiously applying science to their life and production activities, improve efficiency and promote social harmony. In the near future, publication series of CAS Scientific and Cultural Project will constantly come out. I hope that these publications will be welcomed by the reader and that through coordination among CAS science workers, science icons such as Qian Xuesen, Hua Luogeng, Chen Jingrun and Jiang Zhuying, will be more familiar to the public. As a result, the truth-pursuing spirit, rational thinking and scientific ethics will be fully promoted, and the spirit of science workers in courageous exploration and innovation stands eternally in the history of human civilization.

December 2015

Chunli Bai
President of Chinese Academy of Sciences;
Secretary of Leading Party Members' Group
Beijing, China

Preface

A telescope is an instrument which can zoom in distant objects so that people can see clearly. Ordinary optical telescopes are often seen in life and usually consist of objective lens, steering prisms, eyepieces and lens cones. Astronomers use telescopes to observe cosmic space. In 1609, the Italian scientist Galileo Galilei pointed his homemade telescope to the stars for the first time, an unprecedented feat that ushered in a new era of astronomical observation.

Both light and radio signals are electromagnetic waves that propagate at the speed of light, only differing in wavelengths. For thousands of years, man has merely observed the universe through the visible spectrum, while the radiation from celestial bodies covers the entire electromagnetic spectrum. In 1931, Karl Jansky accidentally discovered electromagnetic radiation from the center of the Milky Way. His discovery turned a new page in astronomy, marking the birth of radio astronomy. The emerging discipline contributed to astronomical discoveries in the twentieth century: pulsars, quasars, interstellar and circumstellar molecules and the cosmic microwave background radiation (CMBR), thus becoming the cradle of major discoveries in astronomy. Radio signals from space are extremely weak, so radio telescopes with largest possible apertures are needed to detect more radio signals from celestial objects. However, due to the deformation caused by the weight of telescopes, wind and other factors, the maximum aperture of a conventional trackable telescope can only reach about 100 m.

Twenty-five years ago, a group of Chinese astronomers already envisioned building the "large radio telescope" in China. In 1994, the China Promotion Committee for the Large Radio Telescope was established. Rarely known even by now, the tenacious team brought together over 100 experts from more than 20 universities and institutes across the country to complete the feasibility study of key technologies for large radio telescopes and put forward a preliminary idea for an innovative program to independently build the world's largest single-aperture spherical telescope in China. After that, progress had been made one after another, including site investigation, key technology breakthroughs, project design, construction and acceptance of scaled-down models. Finally, they presented a clear plan for the "Five-hundred-meter Aperture Spherical Radio Telescope (FAST)," which was approved in 2007.

The FAST project is a major national science and technology infrastructure construction project of "the Eleventh Five-Year Plan." It intends to build the world's largest single-aperture radio telescope—the Five-hundred-meter Aperture Spherical Telescope in one of natural karst depressions in Guizhou Province to achieve astronomical observation over a large sky area with high precision. Located in the Dawodang depression, a natural basin in Pingtang County, Guizhou, Southwest China, FAST began construction on March 25, 2016, and was inaugurated on September 25, 2016. Known as the "China Sky Eye," it is a big science facility with independent intellectual property rights and the world's largest and the most sensitive single-aperture radio telescope designed, developed and organized by Chinese scientists.

Throughout the five and a half years, thousands of engineering and scientific personnel, workers and managers have been engaged in this intense, orderly and ingenious construction. They overcame a series of difficulties in Dawodang depression, such as bad weather and harsh environment, and designed and implemented one ingenious process after another to make the FAST project a reality step by step. In the process, more than 20 major contractors have completed systematic tasks including site excavation, ring beams, cable nets, panels, actuators and feed support towers, cable drive system, docking platforms, integrated wiring and electromagnetic compatibility.

This book is a popular science book written by the FAST engineering team, which is organized by Mr. Nan Rendong during his lifetime. The team tries to make people understand what FAST is, what it can do, why it was built and how it was completed. FAST is a powerful instrument for human to explore the universe, offering unprecedented opportunities for new scientific discoveries. If the book attracts more people to understand and love astronomy, then its publication will be more significant.

Gratitude is extended to the editors of Zhejiang Education Publishing House for their valuable comments from the book's planning to finalization, as well as all colleagues who contributed to the planning, writing, finalization and printing of this book.

Finally, we would like to dedicate this book to our esteemed Mr. Rendong Nan.

December 2018   Jun Yan (严俊)
                                                    Manager of the FAST Project

Contents

Towards the Sea of Stars

Haiyan Zhang, Lei Qian, Caihong Sun, Chengmin Zhang, Wenjing Cai, Aiying Zhou, Chengjin Jin, Li Xiao, Dongjun Yu, Qing Zhao, Boqin Zhu, Wenbai Zhu, Lichun Zhu, Ming Zhu, Liqiang Song, Mingchang Wu, Baoqing Zhao, Ming Zhu, Gaofeng Pan, Hui Li, Rui Yao, Youling Yue, Bo Zhang, Rurong Chen, Boyang Liu, Li Yang, Na Liu, Jiatong Xie, Yan Zhu, Hongfei Liu, Zhisheng Gao, and Xiaobing Chen

While looking up to the sky, humans always wondering who we are, where we come from and whether we are alone. In the vast universe, are there other civilizations? For thousands of years, man has merely observed the universe through the visible spectrum, while the radiation from celestial bodies covers the entire electromagnetic spectrum. As radio astronomy observes radio signals from the space, radio telescopes with largest possible apertures are needed to detect more radio signals from celestial objects. Under the mysterious and deep starry sky, FAST serves as a "Sky Eye", leading us to explore wonders and secrets of the universe.

© Zhejiang Education Publishing House 2021
R. Nan (ed.), *The Sky Eye*, China's Big Science Facilities,
https://doi.org/10.1007/978-981-16-3824-4_1

The Shanghai 65 m Radio Telescope ("Tianma"), built in 2012, is the largest fully steerable telescope in Asia

1 Radio Astronomy

Radio and television signals and the well-known visible light are electromagnetic waves. They both travel at the speed of light, but differ in wavelengths. Usually, waves from celestial bodies are called radio waves. Radiation from different celestial bodies covers the entire electromagnetic spectrum, ranging from low-frequency radio to high-energy X-rays and γ-rays (these two rays are also applied in our daily life, such as security scanners and CT imaging).

There are two window frequency bands provided by Earth's atmosphere for humans to observe the universe: visible light and radio waves. Radiation beyond the frequency bands is almost shielded and can only be received outside the globe.

> **Knowledge Link**
>
> **The observation of visible light**
>
> The spectrum of visible light window is 380–780 nm, and the wavelength of yellow-green light in the middle of the spectrum is about 570 nm. At the wavelength, the Sun's radiation is the strongest, and the naked eye is more sensitive to the wavelength. Biologists explain this coincidence from the perspective of theory of evolution. Earth's atmosphere is transparent to electromagnetic waves from several millimeters to dozens of meters, which was found after radio technology was invented. Over the centuries, man merely observed the universe through visible wavelengths, knowing little about the nature of celestial radiation.

In 1931, Karl Jansky, an American radio engineer from Bell Labs in New Jersey who specialized in searching and identifying telephone signal interference, discovered that there was a maximum level of radio interference that occurred every 23 h 56 m 04 s.

After more than a year of precise measurement and careful analysis, in 1932, he formally confirmed this pattern of radio interference was a type of radio emission from outside the Earth's atmosphere and in the direction of the Sagittarius constellation at the Galaxy's center. As a result, man captured radio waves from space for the first time, which turned a new page in astronomy, and people began to study stars and the universe through radio wave observation and thus embraced the birth of radio astronomy. Jansky made his discovery public in 1933 in *Nature*. At that time, he used a rotating antenna array of 30.5 m long and 3.66 m high, detecting a 30-m-wide sectorial beam at the wavelength of 14.6 m (Fig. 1).

Radio astronomy contributed four inspiring astronomical discoveries in the 20th century: pulsars, which confirm the existence of theoretically predicted neutron stars; quasars with small angular apertures similar to stars, emitting more light than galaxies and showing the intense activities of black holes at the center of galaxies; interstellar and circumstellar molecules, which have updated human understanding of complex

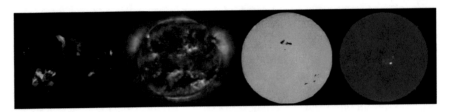

Fig. 1 Images of the sun taken at different electromagnetic wavelengths: X-ray, ultraviolet, visible light, radio (from left to right)

molecules in the interstellar medium; and the cosmic microwave background radiation (CMBR), serving as an evidence for the Big Bang model. Radio astronomy observation brings about six Nobel Prizes in Physics, thus have become the cradle of new ideas and discoveries. These achievements have profoundly affected the human understanding of nature.

Given the development of industrial technology, in particular the progress of high-tech industries such as electronics and computers, and driven by astronomical scientific research, communication industry and national security needs, the detecting capacity of radio astronomy has been enhanced unprecedentedly. Its relative bandwidth (the ratio of the bandwidth to the channel width) exceeds 10,000; its resolution (the ability to observe details of celestial bodies; the higher the resolution, the smaller the angle it can identify) is at least three orders of magnitude greater than all other bands; its detection limit (the ability to detect faint objects) is 0.000 000 000 000 000 000 000 000 000 001 (1×10^{-30}) W/(Hz·m^2). Considering that celestial objects are distant and faint, radio astronomy requires great observation ability.

It is estimated that the amount of celestial radiation energy received by radio telescopes around the world over the past seven decades is insufficient to turn a single page of a book. Reading the books of the universe, therefore, requires radio telescopes with huge apertures.

2 Radio Telescope

Astronomy is an observational science. A telescope is a tool which can zoom in on distant objects so that people can see clearly. Optical telescopes are often seen in life and usually consist of objective lens, steering prisms, eyepieces and lens cones. Astronomers use telescopes to observe cosmic space. It was in 1609, when Italian scientist Galileo Galilei pointed his homemade 32× telescope at the stars, that the mysteries of the universe were first revealed. Optical astronomical telescopes make it possible to observe the Sun, Moon, stars, meteors, comets and so on at great distances from the Earth through the visible spectrum. However, astronomers are not satisfied with optical telescopes, thus turning to radio telescopes.

With the development of radio astronomy, astronomers began to use radio telescopes to observe celestial objects. Unlike optical telescopes, a single-dish radio telescope doesn't have a tall telescope tube, an objective lens, or an eyepiece. It consists of three parts: the reflecting surface which is used to converge the electromagnetic waves radiated by celestial bodies, functioning like a concave mirror that converges the light; the receiver which is used to receive the electromagnetic waves, serving as a radio that receives radio signals; the pointing and tracking device which is used to point the telescope at objects and track them (without regard to the effects of the Earth's rotation).

Radio telescopes are used to observe radio emission from celestial bodies. What are the characteristics of the emission? Primary characteristics are the information of time domain and frequency domain. Time-domain characteristics are the radiation intensity, the variation of radiation intensity with time, etc. Frequency-domain characteristics are the radiation spectrum (i.e., the variation of radiation with frequencies), the variation of spectrum with time, and so on.

Radiation from celestial bodies received by radio telescopes tends to be weaker than artificial signals. The flux of received radiation is generally measured in Jansky; 1 Jansky (Jy) $= 1 \times 10^{-26}$ W/(Hz·m^2).

For a radio source with flux intensity of 1 Jy, its radiation intensity is similar to that of a signal from a mobile phone placed 300–500 km above the ground, where is also the flight altitude of Shenzhou spacecraft's orbital modules.

Knowledge Link

Types of antennas

There are many types of antennas such as array antennas for early radio and television sets, parabolic and reflector antennas for satellite ground stations to communicate with artificial satellites, and radar antennas and cellphone antennas. Almost invariably, these antennas are used to communicate with cooperative targets or to detect the echoes of emitted signals. As a result, there are a variety of designs based on their applications and signal characteristics.

Radio telescope technology is theoretically based on classical electromagnetic theory. From the static electricity and geomagnetism discovered at early time, to the effect of electric current on magnetic needles through the experiment conducted by the Danish scientist Auster in 1820, to the laws of the interaction between electric currents summarized by Ampere, to the law of electromagnetic induction discovered by Faraday, scientific research progress has established a close link between two seemingly unrelated phenomena: electricity and magnetism. The British physicist Maxwell conceived a mechanical model of electromagnetic interaction based on the experimental laws of the interaction between electricity and magnetism, introduced the concept of displacement current as a premise compatible with the law of charge conservation, and proposed basic differential equations linking charges, currents and electric and magnetic fields. These equations, subsequently called Maxwell's

Equations after collation and rewrite, are the culmination of classical electromagnetic theory. They predicted that electromagnetic waves propagate at the speed of light, which was later confirmed experimentally by the German scientist Hertz. Within the framework of classical electromagnetic theory, visible light, ultraviolet light, infrared light, X-rays, gamma rays and radio waves are all electromagnetic waves; only their respective frequencies or wavelengths are different. Led by electromagnetic theory and subsequent experiments, radio-electronic technology has made rapid progress. Representative results are radio communications, radar, radio, television, etc.

Radio-electronic technology, the first technological revolution in human history guided by scientific theories, has greatly improved communication and other technical capabilities and delivered a new perspective for astronomers on the observation of celestial bodies—the electromagnetic spectrum.

After World War II, some radar specialists turned to the study of radio astronomy. With the advancement of electronic technology, since the birth of radio telescopes, telescope specialists have started to enhance their observational capabilities. There are two major directions to improve the performance of radio telescopes: one is to improve the observational sensitivity by increasing the effective receiving area; while the other intends to use multiple antennas to form an antenna array and to improve the observation resolution through enlarging the distance between antennas.

The Effelsberg Telescope in Germany, shown in Fig. 2, was a fully steerable radio telescope built in 1972. With a parabolic diameter (commonly referred to as its aperture) of 100 m, it was proclaimed "the largest machine on the surface of the Earth". An aperture of about 100 m is the ground engineering limit for today's fully steerable radio telescopes. Due to the deformation caused by the weight of telescopes and wind load (the force exerted by the wind on the telescope), fully steerable telescopes with larger apertures are no longer considered by the engineering community. Therefore, is there any way to further increase the reflecting surface aperture? Cornell University built the 1000-foot (305 m) aperture telescope named Arecibo in 1963. Scientists have fixed the spherical crown-shaped reflector in the karst depressions of Puerto Rico, which cannot be rotated (Fig. 3).

On a thousand-ton platform above the reflecting surface, tracking observations of celestial objects in a band of the sky are achieved through the level and pitch motion of the receiver's feed cabin. The telescope is rated as the first of the top 10 projects of the twentieth century, ahead of the Project Apollo, and was used as a set for the Hollywood blockbuster *GoldenEye*.

The reflector of single-dish radio telescopes is often made of continuously laid metal panels, metal mesh, or combinations of the two. Their detection limit is inversely proportional to the square of its aperture (the larger the aperture is, the smaller the detection limit is, i.e., it can detect fainter objects). The resolution is proportional to the aperture (the larger the aperture is, the smaller the objects that can be seen are). The volume of the detected cosmic space and the number of explored objects are proportional to the cube of the aperture. The cost of single-dish radio telescopes with conventional designs is roughly proportional to the cube of the aperture, but as active control technology improves, the aperture-cost relationship is gradually being broken down.

Fig. 2 Fully steerable radio telescope in Effelsberg, Germany

In the 1950s, the radio astronomy community began to develop interferometric arrays and synthetic-aperture radio telescopes, which transmit received signals from several single-dishes with small apertures connected by cables to relevant processing centers. The telescope resolution is proportional to the scale of the antenna array extension (baseline length), while the sensitivity depends on the sum of all single-dish area. Synthetic-aperture radio telescopes are well suited for high-resolution imaging of celestial objects. With advances in time–frequency standards, signal processing, data storage and computer technology, single antennas involved in interferometric observations are no longer tied to physical connections and can be separated by tens of millions of miles, even in conjunction with telescopes in space. The name Very Long Baseline Interferometry (VLBI) is derived from the long observation baseline, which enables VLBI observation to acquire high-resolution details of celestial objects.

The world's most famous synthetic-aperture radio array is the Very Large Array (VLA) at the National Radio Astronomy Observatory (NARO) of the United States.

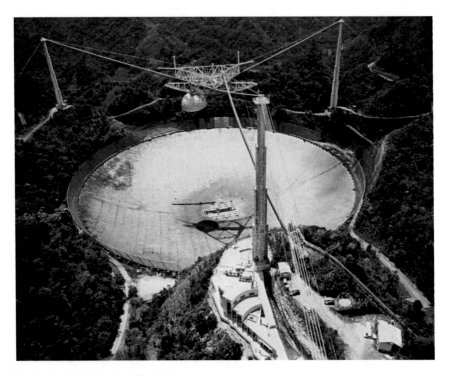

Fig. 3 Arecibo telescope in the U.S

Built in 1981, it comprises 27 dishes, which are 25 m in diameter. In the vast desert of New Mexico, they are arranged in a huge Y pattern up to 36 km across (Fig. 4).

There are many large radio telescope arrays in the world today, such as the Atacama Large Millimeter Array (ALMA) built on the Atacama Plateau in Chile, the Square Kilometer Array (SKA) under construction, etc. Likewise, there are many large single-antennas in operation, under construction or in planning, such as the Shanghai 65 m Radio Telescope ("Tianma") and the U.S. Green Bank Radio Telescope in operation, the Xinjiang Qitai 110 m Radio Telescope (QTT) in planning, etc. Each of them has its own scientific objectives, and it is difficult to distinguish them in terms of overall performance. Single-dish telescopes are easy to develop competitive edge when it comes to sensitivity, wavelength, and relatively long service time. Their scientific advantage lies in discovery, while that of the radio telescope array is that it does fine imaging of celestial objects. There are no clear differences between them. Single-aperture telescopes are basic units of telescope arrays, whose phased array observation mode is similar to that of single-aperture telescopes in terms of the data structure.

As an international collaboration, ALMA is a large millimeter/submillimeter radio array. Consisting of sixty-four 12-m dishes, it was jointly funded by the European Southern Observatory (ESO) and the National Radio Astronomy Observatory

Fig. 4 The VLA in the U.S

(NRAO) of the United States, in cooperation with the National Research Council Canada (NRC) and with free use of land authorized by the Government of Chile.

The one comparable with ALMA is the Atacama Compact Array (ACA) invested by the National Astronomical Observatory of Japan (NAOJ), which consists of twelve 7-m and four 12-m dishes. As a component of ALMA, it enhances the capacity of detecting extended surface sources. ALMA has a maximum baseline of 14 km, an operating frequency of 30–950 GHz, a resolution of 10 milliarcseconds and a continuous spectral sensitivity of several microjanskys. These characteristics make ALMA an unparalleled instrument for the cold universe imaging and spectroscopic observation. Its main observational objectives are the solar-terrestrial scale structure of gas and dust clouds, planets and other thermal-emitting celestial bodies; the kinematics of protoplanetary disks; the chemical composition of galactic molecular clouds; the kinematics of optical Active Galactic Nuclei (AGNs) and quasars; thermal imaging of asteroids, comets and Kepler belt objects; the photospheric structure of neighboring giants; the chemical composition, structure and gas dynamics of merging galaxies; and the imaging and dynamics of molecular gas in distant galaxies. Compared with other radio astronomy equipment, ALMA focuses more on thermal emission, astrochemistry and cosmic life environment (Fig. 5).

The Square Kilometer Array (SKA) is a multinational project for a giant radio telescope which was initially conceived in 1991, and in 1993, The International Union of Radio Science (Union Radio-Scientifique Internationale, URSI) formally proposed and immediately established a working group to coordinate international cooperative research. In 2006, scientists made a preliminary selection of different unit

Fig. 5 ALMA in chile

technology concepts and determined two candidate sites. The SKA includes hundreds of stations with equivalent apertures of 100–200 m, with a total receiving area of about 1 km². 50% of the receiving area is in the central 5 km region to improve the capacity to monitor low-brightness sources, with a maximum baseline of more than 3000 km to get the high-resolution images of stars, the interstellar medium and distant galaxies. It operates in the 100 MHz–25 GHz frequency range. Based on current conceptual studies, it is estimated that it would take 3 arrays to implement such a frequency coverage: The low frequency array consists of electrically-scanned phased array feeds (PAFs). Both the intermediate frequency (IF) array and the high frequency (HF) array use small-aperture fully steerable dishes with focal plane arrays (FPAs) feeds and broadband feeds respectively. SKA requires a large field of view (FoV) to improve inspection efficiency, and 1 square degree is needed for the half-power beam width (HPBW) FoV in the L-band, and 200 square degrees for low frequencies. Inexpensive small-antenna technology and the development of new feeds are vital to the SKA (Fig. 6).

The Shanghai 65 m Radio Telescope ("Tianma") project was officially launched in February 2009 and completed in 2012.

The radio telescope adopts a modified Cassegrain reflector which consists of two reflecting surfaces: a 65-m-diameter primary reflecting surface, the f-number of which is 0.32, and a 6.5-m-diameter secondary reflecting surface. Its main reflecting surface area is about 3780 m², equivalent to the size of 9 standard basketball courts. It functions like a huge "ear", clearly "hearing" faint radio signals from deep universe. Shanghai 65 m Radio Telescope can rotate more than 360 degrees in the horizontal direction to point to different directions, while in the elevation direction it varies about 90°. It is the largest fully steerable radio telescope in Asia. The telescope is well suited to perform the VLBI orbiting and positioning for the lunar exploration phase II and III, carry out future deep space exploration, and substantially contribute to the study of astronomy (Fig. 7).

The Green Bank Telescope (GBT) is a 100-m single-dish telescope with an actual aperture of 100 m × 110 m and a total mass of 7300 tons, making it the largest fully

Fig. 6 The effect drawing of the SKA

steerable radio telescope in the world. The virtual rotating parabola of the primary mirror is 208 m, cut from 4 m away from the rotation center, and the quadric is suspended on a cantilever extending from the edge. Due to the deflected illumination, the electromagnetic waves reach the full mirror directly without being blocked by the feed support, improving the telescope's efficiency while avoiding structural scattering. It has 2000 panels, each with actuators at the corners. Laser measurement equipment is installed between the telescope structures and on the ground to monitor in real time the deformation of the system caused by the weight of telescopes, wind load and temperature and feed back to the actuator to compensate.

The GBT saw "first light" (that is, detected its first radio waves from space) in 2000, and was officially put into operation in 2003. It operates in the 290 MHz–52 GHz frequency range and is expected to rise to 95 GHz in the future. To ensure the service time of the telescope, scientists have established the National Radio Quiet Zone.

Since its official operation, the GBT has yielded first-class scientific results in many fields such as pulsars and interstellar chemistry. It is involved in the monitoring of the sole binary pulsar, PSR 10,737-3039AB, and has discovered 21 ms pulsars in globular clusters, a giant radio lobe at the Galaxy center, possible traces of ice water in lunar poles in collaboration with the Arecibo Telescope, and two new interstellar molecules at the center of the Galaxy.

China's FAST is currently the world's largest single-dish radio telescope in terms of aperture, with the advantage of a huge receiving area. Naturally, FAST will form an antenna array with other antennas to enhance the array's observational sensitivity (Fig. 8).

Fig. 7 Shanghai 65 m radio telescope ("Tianma")

Knowledge Link

The service time of FAST

The vast majority of radio telescopes in operation around the world should have been out of service, such as the U.S. Arecibo telescope, which has been in service for more than 50 years. The design life of FAST is 30 years. With some modifications and upgrades, it still can be in service after the expiry date.

Fig. 8 The GBT in the U.S

3 Why We Need Large Aperture Radio Telescopes?

Many technical specifications improve exponentially over time. For example, according to Moore's law of advances in large-scale semiconductor chips, a computer's integrated computing power doubles every 18 months. The Living Stone curve, which depicts the progress in the sensitivity of radio telescopes, shows an exponential increase in sensitivity from 1940 to about 100,000 times by 2000, doubling every three years.

The construction of large telescopes is not driven by economic interests, but by human impulse to create and explore, primarily to address the hot issues at the frontiers of astronomy. They are built and run with discoveries and breakthroughs. The 76.2 m Lovell Radio Telescope from the Jodrell Bank Observatory discovered gravitational lenses; the Parkes 64 m Radio Telescope in Australia discovered quasars; the Westerbork Synthesis Radio Telescope (WSRT) in Netherlands discovered the largest radio galaxy; the Very Large Array in the United States saw the center of the Milky Way at optical wavelengths obscured by dust for the first time; the Arecibo's discovery of pulsed binaries provided evidence for the existence of gravitational waves. Generally speaking, larger telescopes means more scientific output, making it possible to discover new celestial objects and see deeper into the universe (Fig. 9).

Certainly, nothing is absolute. New astronomical discoveries require good equipment, as much as they require the ingenuity and insight of scientists, and perhaps good luck. But large-aperture telescopes can detect faint celestial objects, capture transient phenomena and probe deeper into space. Large aperture telescopes can provide more and better statistical samples of observations for scientists to discover and refine patterns, and detect more exotic celestial objects, providing opportunities for scientific breakthroughs. Scientific prediction is risky. When Columbus was given "project funding" to raise a fleet, he was rewarded with ships full of gold, silver, spices and a new continent. Both Queen Isabella and Columbus had no idea that there was a new world, but Queen Isabella was wise and virtuous because she knew that vessels were suitable for long-distance voyages. As predicted by the Livingstone Curve, in

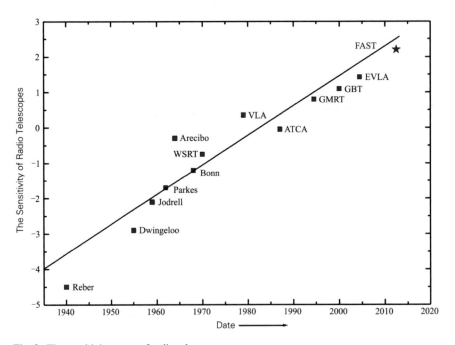

Fig. 9 The sensitivity curve of radio telescopes

the current decade there should be a radio telescope with a detection capability of approximately 500 m. The FAST project conforms to the prediction as well as the development trend.

Knowledge Link
Is FAST the largest radio telescope?
 FAST is the world's largest single-dish radio telescope, but not the largest radio telescope. Based at an altitude of 2100 m in the North Caucasus, the world's largest radio telescope is the Russian RATAN-600 Radio Telescope, a very rare 605 m banded radio telescope.

The "China Sky Eye" Explores the Universe

Haiyan Zhang, Lei Qian, Caihong Sun, Chengmin Zhang, Wenjing Cai, Aiying Zhou, Chengjin Jin, Li Xiao, Dongjun Yu, Qing Zhao, Boqin Zhu, Wenbai Zhu, Lichun Zhu, Ming Zhu, Liqiang Song, Mingchang Wu, Baoqing Zhao, Ming Zhu, Gaofeng Pan, Hui Li, Rui Yao, Youling Yue, Bo Zhang, Rurong Chen, Boyang Liu, Li Yang, Na Liu, Jiatong Xie, Yan Zhu, Hongfei Liu, Zhis heng Gao, and Xiaobing Chen

The Five-hundred-meter Aperture Spherical Telescope, or the FAST for short, known as the "China Sky Eye", is the world's largest and most sensitive single aperture radio telescope. It is located in a karst depression in Pingtang County, Qiannan Buyei and Miao Autonomous Prefecture, Guizhou Province. The main construction includes site investigation and excavation, the active reflector, feed support, measurement and control, receiver and terminal, and observation base. The FAST will achieve the goal of astronomical observation over a large sky area with high precision.

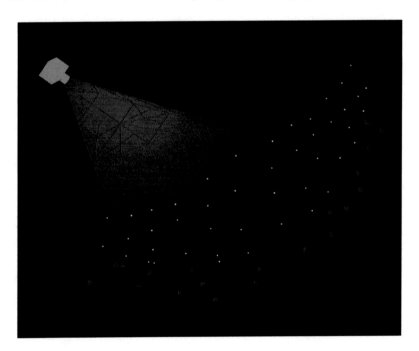

The crystalline lens of the "China Sky Eye"—the active reflector

© Zhejiang Education Publishing House 2021
R. Nan (ed.), *The Sky Eye*, China's Big Science Facilities,
https://doi.org/10.1007/978-981-16-3824-4_2

The FAST project is a major national science and technology infrastructure construction project of China's "the Eleventh Five-Year Plan". In this project, the natural karst depression in Guizhou is chosen as a telescope site to build the world's first single aperture radio telescope, the Five-hundred-meter Aperture Spherical Telescope, in order to realize astronomical observations in a large celestial area with high precision. The Chinese Academy of Sciences is the construction authority of the project, the Government of Guizhou Province is the co-construction authority, and the National Astronomical Observatory of the Chinese Academy of Sciences is the legal entity. Construction of the telescope began on March 25, 2011, and was completed and inaugurated on September 25, 2016. Located in the Dawodang depression of Jinke Village, Kedu Town, Pingtang County, Qiannan Buyei and Miao Autonomous Prefecture, Guizhou Province, FAST is the world's largest and the most sensitive single-dish radio telescope with independent intellectual property rights in China.

FAST is innovatively designed, developed, manufactured and organized by Chinese scientists. Its main objectives are: to lay a 500-m-diameter spherical active reflector in the karst depression in Guizhou, and to form a 300-m aperture instantaneous parabolic surface in the observation direction through active deformation control; to enable high precision point tracking without rigid connection between the feed and the reflector by adopting a platform of cable-supported light-weight feed source of the optical, mechanical and electronic integration, together with the secondary adjustment device in the feed cabin; to configure a multi-band, multi-beam feed and receiver system covering 70 MHz–3 GHz in the feed cabin; to develop different end-use devices for the FAST science objectives; and to build a world-class astronomical observatory.

In the design and construction process, FAST achieved three independent innovations: first, the natural karst depressions in Guizhou were used as the site; second, thousands of units were laid in depressions to form a 500-m spherical active reflector, which formed a 300-m aperture instantaneous parabola in the direction of the radio source, enabling the telescope receiver to be at the focal point like traditional parabolic antennas; third, a light cable-driving mechanism and parallel robots were used to achieve high-precision positioning of the receiver.

1   What Can the FAST Do?

FAST aims to cover a wide range of astronomical topics: initial turbidity of the universe, dark matter, dark energy and large-scale structure, the evolution of galaxies and the Milky Way, stellar-like objects, and the planets of our solar system and adjacent space. FAST intends to answer scientific questions about astronomy, humans and nature. Its potential scientific output is still hard to predict today.

1.1 Space Beacons—fSearch and Timing Observation of Pulsars

Half a century ago, the discovery of pulsars was one of the greatest achievements in radio astronomy. In 1967, under the guidance of her mentor Antony Hewish, Jocelyn Bell, a postgraduate at Cambridge University, conducted radio observation for quasi-periodic photometric variation in small-diameter radiation sources due to interplanetary scintillation. She accidentally discovered a radio source with a stable period of 1.33 s, but a single pulse width lasting only 0.04 s.

Astronomers at the time did not know much about the properties of the radio source. Therefore, she once thought it was artificial signal interference, and even ventured to speculate that it was a signal sent to Earth by extra-terrestrial civilizations.

In the following year, Hewish and others finally determined that the pulsed radio source should come from rapidly-rotating compact objects outside the solar system, especially neutron stars. The name "pulsar" for this kind of radio sources was also officially introduced in 1968.

Figure 1 shows the multi-band pulse frequency-phase diagram of the first pulsar recorded by FAST during testing. From bottom to top, frequencies vary from low to high, and pulses arrive earlier. The signal–noise ratio is over 5000 for this observation, which is a testament to the telescope's ultra-high sensitivity (Fig. 2).

Neutron stars are produced when massive stars (8–25 times the mass of the Sun) exhaust the fusion fuel in the central region and explode as supernovae at the end of their lives. During the eruption, the star's outer envelope was thrown outward, interacting with the interstellar medium to produce supernova remnants, while the core lost the outward pressure provided by the nuclear reaction and collapsed inward under gravitational forces.

For stars more than eight times the mass of the Sun, their central masses exceed the Chandrasekhar limit (approximately 1.4 times the mass of the Sun) that allows for the stable existence of white dwarfs. Neutron stars are only dozens of kilometers in diameter, their cross-sectional area is tantamount to the size of a city, and their total mass is in the same order of magnitude as the mass of the Sun, so their density is extremely high, up to hundreds of millions of tons per cubic centimeter, and a ping-pong ball-sized neutron star is almost as massive as an entire mountain range on Earth. Since neutron stars at birth retain most of the angular momentum of their predecessor—fixed stars, but on a much narrower scale, they usually have a faster rotation, and the short pulse period directly reflects their rotation period. At the same time, due to the conservation of magnetic flux, the surface area of collapsed stars is greatly reduced while the magnetic field surges. The typical magnetic field of a neutron star can reach up to 10^{12} gauss and even more, while that of the Earth is only about 1 gauss, and the typical intensity of the magnetic field in the solar active region is only a few thousand gauss.

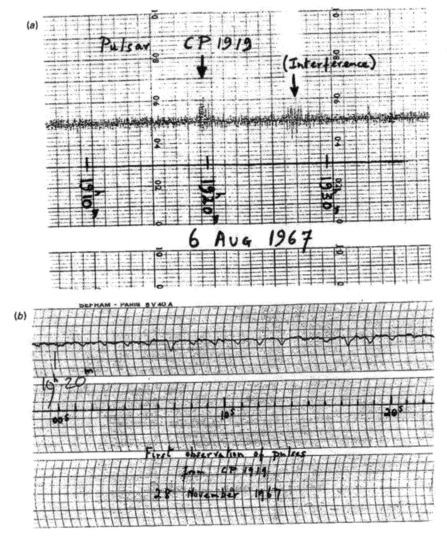

Fig. 1 Chart recordings of PSR B1919 + 21, the first discovered pulsar. *Credit* Lyne and Graham-Smith (2012)

Knowledge Link

Chandrasekhar limit

   The maximum mass that a non-rotating star can resist gravitational collapse through electron degeneracy pressure. The currently accepted value of the Chandrasekhar limit is about 1.4 M⊙ (the solar mass). Stars with masses

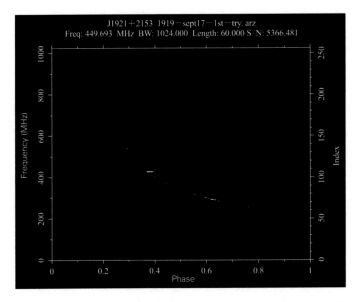

Fig. 2 Multi-band pulse frequency-phase diagram of the first pulsar recorded by FAST

less than this limit are prevented by the electron degeneracy pressure from collapsing, developing into white dwarfs, while those with masses above this limit are subject to further collapse, evolving into neutron stars or black holes.

Electron degeneracy pressure

The Pauli exclusion principle disallows two identical half-integer spin particles (electrons and all other fermions) from simultaneously occupying the same quantum state, resulting in a particular manifestation of quantum degeneracy pressure.

The discovery by Bell and her mentor fills the gap of human's perception of life cycles of stars, especially those of massive ones, thus having far-reaching theoretical implications. Because of his work on pulsars, Huyghe was awarded the 1974 Nobel Prize in Physics.

Nonetheless, the first discoverer, Bell was not one of the recipients of the prize (Fig. 3).

Neutron stars are not necessarily detectable as pulsars on Earth. One requisite for telescopes to receive pulses is that radio beams from stars have to periodically sweep across the Earth. According to the prevailing theory, the radio beams are emitted from the polar-crown region in the magnetic field of neutron stars and are driven by their rotational energy. Usually, the magnetic axis of a neutron star does not coincide exactly with its rotation axis, but leaves an inclination. Radio beams rotate around

Fig. 3 A comparison between the size of a neutron star and that of Manhattan. In contrast, small to medium-mass stars like the Sun can only quietly evolve into white dwarfs which are similar to Earth in size at the end of their lives. *Credit* NASA/Goddard Space Flight Center

the rotation axis, sweeping past the Earth to produce the pulse profile we receive. This is the beacon model of pulsars (Fig. 4).

As releasing radiant energy, neutron stars' rotation gradually slows down, and eventually the luminosity of radio beams drops below the telescope's detection threshold. Therefore, it seems that isolated neutron stars can only be observed when they are young, and their existence as pulsars is no more than a million years, which is not very long compared with the period of stellar evolution (Fig. 5).

It is estimated that the total number of pulsars in the Milky Way may be around 100,000, but only over 2700 have been recorded. The vast majority of them are radio pulsars, with another 200 or so emitting X-rays and gamma rays, while only a small fraction of them emit high-energy pulses and faint radio sources. There are no more than 20 optical pulsars known to us, with individual examples such as pulsars at the center of the Crab Nebula emitting full-band pulses. As a result, scientists rely heavily on radio astronomy observation to search for more pulsars and monitor known pulsars over long periods of time, and related topics are also among the important scientific research goals of FAST once it is fully operational.

What is the point of searching for new pulsars while continuously monitoring known pulsars? This has to be answered from various aspects. Firstly, these objects are extremely dense and have immensely strong magnetic and gravitational fields, so they can serve as an excellent natural laboratory for researchers to explore extreme states of matter, high-energy astrophysical processes, and relativistic effects.

The shortest rotation period of pulsars that have been discovered is 1.39 ms. If sub-millisecond pulsars can be detected in the future, it will effectively support the hypothesis that pulsars are quark stars; if we can't detect any sub-millisecond pulsar,

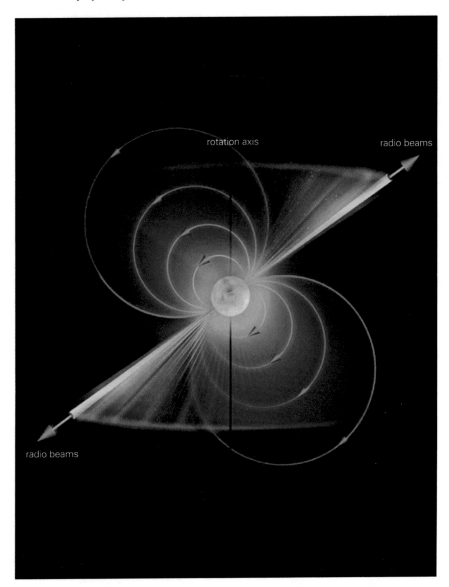

Fig. 4 The structure of pulsars. The blue part in the figure represents the magnetic induction lines, the yellow part stands for the radio beams, and the red line refers to the star's rotation axis. *Credit* Bill Saxton, NRAO/AUI/NSF

Fig. 5 The appearance of the Crab Nebula (M1). It's the remnant of SN 1054, which exploded in 1054. The pulsar at the center of M1 is one of the brightest radio sources in the sky, and the only one whose age can be exactly determined so far. The supernova explosion was recorded by many countries in the northern hemisphere. The most detailed records are found in China's Song dynasty's history books. *Credit* ESO

only to find more pulsars with rotation periods of 1 ms or more, the theory of the physical state under extreme conditions can also be constrained to a large extent by analyzing their periodic distribution patterns.

In addition, pulsars have a strong gravitational field because of their high density and small size, contributing to the thriving gravitational wave detection. Gravitational waves are ripples in spacetime predicted by Einstein's theory of general relativity, and the first indirect test was with the aid of the binary pulsar named PSR B1913 + 16, whose discoverers, Russell Hulse and Joseph Taylor, earned the 1993 Nobel Prize in Physics. Based on the theory of relativity, as stars in a binary pulsar system orbit around each other, some orbital energy would be converted to gravitational waves, causing orbital decay. Observations show that the system's orbital variation over several decades are in perfect line with the theoretical predictions. Subsequent discoveries of more binary pulsar systems, especially the evolution of the binary pulsar PSR J0737-3039, have reached the same conclusion. More binary pulsars are desirable in order to take full use of orbital decay to test the theory of relativity.

Verification of gravitational waves through orbital decay of binary pulsars presupposes precise measurements of the time it takes for the pulses to reach Earth, revealing tiny orbital variation. The orbital variation of PSRB 1913 + 16 over the years are measured in centimeters, as is the case for most other systems. The good news is that pulsars' great rotational inertia, stable rotation, and regular periodic variation ensure their excellent punctuality, making it possible to conduct high-precision observation. For isolated pulsars without companions, another application of precise timing is to directly detect gravitational waves—forming pulse timing arrays with a large number of pulsars all around the universe, and finding the space–time perturbations, or gravitational wave signals, that spread all over the universe by comparing the small but regular correlation changes in relative standard values of the time of arrival (ToA) of pulses from different stars.

Given that gravitational waves cause space–time relaxation, as they pass through the Earth, the distance between the Earth and pulsars is slightly changed, at which point the predicted values of relativity and the ToA of pulsars vary in nanoseconds. The technique is suitable for finding low-frequency (10^{-9}–10^{-6} Hz) gravitational waves generated by certain processes such as the massive black holes' orbit around each other. However, detecting the frequency band is beyond the capacity of ground-based gravitational wave detectors such as the Laser Interferometric Gravitational-Wave Observatory (LIGO), the Virgo interferometer of the European Gravitational Observatory (VIRGO), and even the Laser Interferometer Space Antenna (LISA) under construction. The largest current pulsar timing array, consisting of dozens of millisecond pulsars, has not yet detected gravitational wave signals for sure, but the upper limit of the gravitational wave background it provided is already superior to other means such as the Planck satellite. If we can expand the size of the array, or increase the measurement sensitivity while further extending the observation time, it is expected that new ways of detecting gravitational waves will be available in the near future, enabling us to understand the life and death of supermassive black holes at the center of distant galaxies by treating pulsars as a medium (Fig. 6).

Knowledge Link

The detection of gravitational waves

On August 17, 2017, the LIGO and the VIRGO detected for the first time the merger of two neutron stars in the galaxy NGC 4993 40 megaparsecs (130 million light years) away. The merger was named GW170817, which produced not only gravitational wave radiation, but also electromagnetic radiation, and a sub-gamma-ray burst occurred two seconds after the merger. The difference between a binary neutron star merger and a black hole merger event is that the signal duration of gravitational waves triggered by the black hole merger is very short, usually only a second or less, but the signal triggered by a neutron star merger can last up to a minute. This is because neutron stars are less

Fig. 6 Ripples in Spacetime caused by gravitational waves affect pulsar signals transmission. *Credit* David Champion, Max Planck Institute for Radio Astronomy

massive than black holes and the intensity of the gravitational waves generated by mergers is lower. Therefore, it takes longer for their orbital decay and merge. Neuron stars' longer signal duration enables researchers to test Einstein's theory of general relativity more precisely, and perhaps gives us more clues to the origin of neutron stars. The observation of short gamma-ray bursts is also significant, and its association with gravitational waves has been confirmed for decades. Moreover, the discovery of binary neuron star mergers enables the multi-messenger astronomy enter into a new stage.

The 2017 Nobel Prize in Physics was divided, with one half awarded to Rainer Weiss, professor of MIT, and the other half jointly to Barry C. Barish and Kip S. Thorne, both are professors of CIT, for decisive contributions to the LIGO detector and the observation of gravitational waves.

For pulsars emitting high-energy radiation rather than coordinated radio signals, their origin and evolution still remain to be explained. For instance, is the radio faint because it is the case at a certain evolutionary stage of pulsars, or just a geometric effect due to the fact that the high-energy radiation beam is wider than the radio radiation beam? Interconversions between radio pulsars and high-energy pulsars have been found. Are those common? What is the mechanism beneath them? We have currently witnessed such high-low energy conversions in millisecond pulsar members of three binary systems (e.g., PSR J1023 + 0038). This phenomenon is interpreted as the accretion process of pulsars modulated by pulsar winds: matters from companion stars accrete to pulsars. The binary stars are close enough that

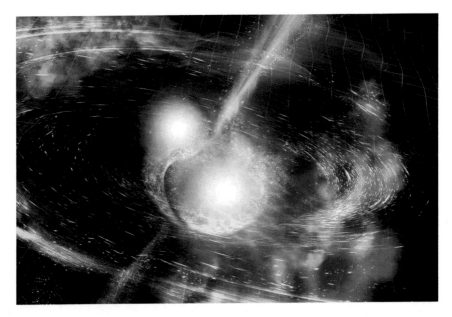

Fig. 7 Binary neutron star merger. *Credit Science* in 2017

during the flow of matters from the low-mass companion star to the pulsar, the pulsar's rapid rotation and pulsar winds generated by the strong magnetic field largely block accretion streams, preventing these streams from approaching the pulsar, while exposing radio radiation beams; when suction flows break through the blockage and surge near the pulsar, they form an incandescent suction disk that turns on the high-energy radiation but shuts off or obscures the radio radiation. Nonetheless, do other millisecond pulsars, and even ordinary pulsars possess similar conversions? Why? To answer these questions, it is necessary to detect more pulsars with radiation conversion (Fig. 7).

Of course, as an outcome of supernova outbursts, the relationship between pulsars and supernova relics is also a topic worth discussing. Researchers have only found pulsars in about 100 supernova remnants, including the Crab Nebula created by the SN 1054 explosion. There are relatively few coordinated instances known, because the lifetime of supernova relics, typically 100,000 years, is much shorter than that of typical pulsars, and because the pulsar's magnetic field (and then radiation beams) must be properly directed to be received by observers on Earth. Besides, the impact on pulsars at the beginning of their formation tends to cause them to move sideways, and 100,000 years after their birth, pulsars are likely to have moved out of the range of the original supernova relics, or at least significantly deviated from the center of relics, which invariably makes the search more difficult. The discovery of more pulsars coordinated with supernova remnants and a census of the stars' spatial motions would undoubtedly deepen the understanding of the end-of-life processes of massive stars (Fig. 8).

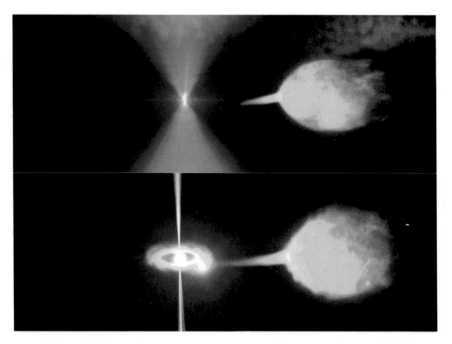

Fig. 8 The conversion process between the radio state (upper) and the high-energy state (lower) of the transformable pulsar PSR J1023 + 0038. The pulsar emits radio waves (green) when accretion streams from the companion star are blocked by pulsar winds, while once the streams reach the surface of the star, they form an accretion disk and induce high-energy radiation (purple). *Credit* NASA's Goddard Space Flight Center

Knowledge Link

Coordination

It refers to celestial objects that appear to be in the same direction (same position in the sky), but not necessarily at the same distance. Or it is highly possible that they are the same object. It can also be interpreted as the consistency between the observation and the theory.

Compared with the states of matter inside compact stars and extreme astrophysical processes, equally important is the study of the action of pulsars on the interstellar medium. Since radio emission emitted by pulsars disperses and scatters as it propagates in the interstellar cold plasma medium, these effects can be applied to probe the distribution of free electrons in the Milky Way, helping us better understand the nature of ionized matter. Therefore, the multi-band monitoring of a large number of pulsars as samples can enhance our understanding of the interstellar medium, and the results are favorable for broader research fields, such as exploring the structure of the Milky Way (Fig. 9).

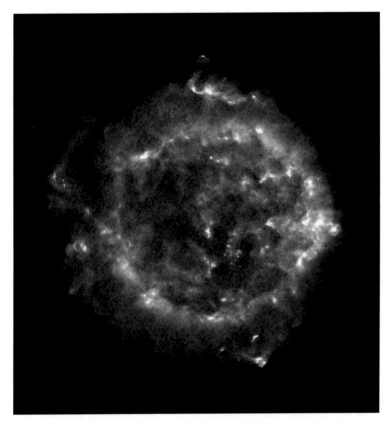

Fig. 9 Radio image of the Cassiopeia A, a supernova remnant. The supernova remnant is the brightest radio source in the universe, detectable at all wavelengths. Its bright crustal structure results from the interaction of supernova ejecta with the interstellar medium. *Credit* NRAO/AUI

Surely, the precise observation of pulsars is of great benefit to basic scientific research, national production and our daily life.

The high stability of a pulsar's rotation is comparable to that of the most sophisticated atomic clocks, especially millisecond pulsars, whose pulse error is so small that they are known as "natural Cesium atomic clocks". Therefore, if this can be used in navigation, the receiver can be precisely located based on triangulation by comparing received pulsar signals and precisely-measured pulsar periods with coordinates. Although the technology is still in the initial stage, it enjoys a bright future. This is because that the biggest advantage of pulsar navigation over common navigation methods is that it's not as dependent on artificial satellites or weather as traditional astronomical navigation. It is more widely applicable, and still available when satellites are malfunctioning. Furthermore, pulsar navigation is more convenient for interplanetary probes which are so far from Earth that cannot rely on navigation satellites.

Up to now, you should have had a general idea of the significance of pulsar observation as well as the existing difficulties. Compared with telescopes in existence, the FAST's major strengths are its extensive reception area and broad sky coverage. With a 300-m effective illuminated aperture in the 500-m-diameter reflector, it is needless to say that FAST is among the best globally. With its innovative active reflector and feed cabin, FAST has a wider field of view than fixed radio telescopes such as the Arecibo Observatory, viewing objects within 40° of the zenith. Located in the Dawodang depression, a natural basin in Pingtang County, Guizhou, the telescope covers the declination of $-14°4'$ to $+65°6'$ in the sky. In contrast, the Arecibo Observatory can only observe objects within 19.7° of the zenith, barely able to view objects appearing in the Southern Hemisphere.

With its large receiving area, wide field of view, new receiver system, and the latest programs developed by the team, FAST can make a huge difference in the pulsar research. The telescope can be 3–5 times more sensitive than previous instruments with the best performance. Since pulsars in the Milky Way approximately scatter around the galactic disc, at least half of them can be detected by FAST. It is theoretically estimated that around 4,000 pulsars of different kinds will be detected once the telescope completes its first year of sky survey, more than the total number of known pulsars.

Nonetheless, the high detection rate means more than the increase in the number of pulsars.

First, the high sensitivity enables FAST to detect older pulsars with longer, fainter periods that previous telescopes are unable to detect. Second, it contributes to the monitoring of known pulsars, enabling FAST to make achievements in observing that should be made by other telescopes for decades.

Theoretical estimates suggest that FAST can observe the pulse period variation of pulsars over a shorter period of time because it is able to acquire the pulse profiles of pulsars with shorter integration time, thus assisting scientists in better understanding the timing residuals caused by the nature of pulsars, and systematic errors in the detection of gravitational waves with pulsar timing arrays in the future. It is estimated that FAST's sensitivity of 20-pulsar timing observation is 10–16 through 10-year observation, setting the optimal limit on this band so far. In addition, FAST is expected to discover some new types of pulsars. New search programs enable the telescope to detect beyond the known "main sequence" band of pulsars in the period-period derivative diagram and search for pulsars in a much wider parameter space, filling in possible gaps (Fig. 10).

The following are the FAST possible discoveries: binary systems consisting of neutron stars and black holes, sub-millisecond pulsars with periods of less than a millisecond, the first radio pulsar in anagalactic nebulae beyond the Large and Small Magellanic Clouds, and more pulsars with radiation beams converting between the high energy and the radio. Such discoveries, once made, would greatly advance theories and applications relevant to pulsars.

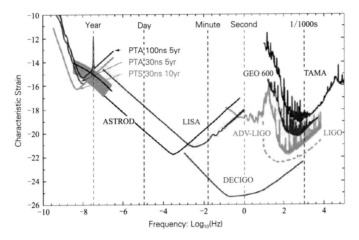

Fig. 10 The sensitivity curves of gravitational waves detectors. The three lines on the upper left of the figure are the FAST sensitivity limits for the timing observation of 20 ms pulsars. ASTROD: astrodynamical space test of relativity using optical devices. LISA: laser interferometer space antenna, a space gravitational-wave observatory; DECIGO: deci-hertz interferometer gravitational-wave observatory; LIGO: laser interferometer gravitational-wave observatory; ADV-LIGO: advanced LIGO, the second-generation ground-based laser interferometry experimental facility. GEO 600: a gravitational wave detector and laser interferometer with 600-m-long arms; TAMA: a gravitational wave detector and laser interferometer with 300-m-long arms. *Credit* Nan et al. (2011)

1.2 The Probe of Cosmic Evolution—Neutral Hydrogen

Hydrogen is the oldest, simplest and most abundant chemical substance in the universe, accounting for 76% of all baryonic mass and 92% of the number of atoms. Hydrogen exists in the universe in various forms, either as hydrogen atoms, hydrogen ions, hydrogen molecules, or as molecules with other elements. Among them, neutral hydrogen atoms are abundant in the universe and ubiquitous even in low-density environments. The ground state of neutral hydrogen consists of an electron bound to a proton. The 1420 MHz (21 cm wavelength) radiation comes from the transition between the two levels of the hydrogen 1 s ground state, slightly split by the interaction between the electron spin and the proton spin. The splitting is known as hyperfine structure. The spectral line was first predicted by Dutch physicist Van de Hulst in 1944, first detected in the Milky Way by Ewen and Purcell at Harvard University in 1951, and successfully confirmed by Dutch and Australian scientists in the same year. Since then, the observation of the Hydrogen 21-cm Line has been important to the spectral line study.

The comprehensive and detailed map of neutral hydrogen in the Milky Way, the spiral galaxy in which we live, has an unparalleled advantage over other galaxies in studying the properties of the interstellar medium. Even for the nearest galaxy, M31, l arcmin beam of the telescope corresponds to a 200-parsec line scale and even the farthest region of the Milky Way corresponds to details 20 times larger than this resolution. The 21 cm radiation of neutral hydrogen superimposes regions of various scales, temperatures and densities in this direction, carrying information on the physical state of the interstellar medium in different environments and on the interactions between the various stable phases. The spatial distribution of neutral hydrogen intensities and velocities can not only obtain the disc and spiral arm structure of the Milky Way, and thus deduce the structure, formation and evolution of spiral galaxies, but also provide information on galactic dynamics to test dark matter models.

Most of the data we know so far about the large-scale distribution of neutral hydrogen in the Milky Way mainly comes from the Leiden-Argentine-Bonn (LAB), an all-sky survey project. The survey was made by Dutch and Argentinean scientists in the 1980s through a 25-m and a 30-m radio telescope. The observation resolution is rather low, about half a degree, with a sensitivity of 0.09 K. In October 2016, new single-dish all-sky neutral hydrogen surveys, the Effelsberg-Bonn HI Survey (EBHIS) in the Northern Hemisphere and the Parkes Galactic All-Sky Survey (GASS) in the Southern Hemisphere, were merged through the HI4PI project to form the latest map of neutral hydrogen in the Milky Way. With the resolution of 9 arcmin and 16 arcmin for their survey observation respectively, this project will significantly improve our knowledge of the large-scale distribution of neutral hydrogen in the Milky Way (Fig. 11).

The Milky Way, like most spiral galaxies, has differential rotation, with the rotation curve flattening out within 5–27 kiloparsecs.

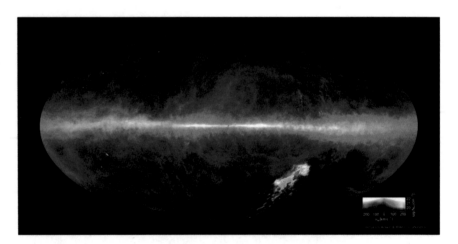

Fig. 11 The all-sky map of neutral hydrogen in the Milky Way drawn by the HI4PI project. The data comes from the Effelsberg 100-m Radio Telescope in Germany and the Parkes 18-m telescope in Australia. In the figure, the blue part represents that neutral hydrogen is running towards us at a lower radial velocity, while the green part stands for that neutral hydrogen is moving away from us at a lower radial velocity. The bright band in the center is the galactic disc, and the bright spots in the lower right are the adjacent large and small Magellanic clouds. *Credit* NASA

Knowledge Link

Differential rotation

This means that points further from the rotation center will travel at greater speeds than those closer in.

Kiloparsec

It's a unit of length in astronomy. 1 kiloparsec = *1000 parsecs = 3260 light years. A light year is the distance that light travels in vacuum in one Julian year, 1 light year = 9,460,730,472,580,800 m (about 9.46 × 10¹² km).*

The fundamental constants currently accepted by the International Astronomical Union are: 8.5 kiloparsecs, the distance between the Sun and the center of the Milky Way system, and 220 km/s, the velocity of the Sun orbiting around the Galactic Center. It's known that neutral hydrogen atoms in the Milky Way form a thin disc of gas along the galactic plane, centered at the heart of the Galaxy. Its scale is about three times the size of a stellar disc, filled with shell, spine and chimney structures at all scales, which are closely related with the life and death of fixed stars in stellar discs, demonstrating the "ecosystem" in the Galaxy. Due to the gravitational pull of the Large and Small Magellanic Clouds, the gas disc is distorted, with the north side being larger and the south side smaller. The distortion is seen in molecular clouds, stars, ionized hydrogen areas, and other galactic disc tracers. Within 0.7 solar radius of the galactic disc core, the gas disc is flat and broad, with a height of about 220 parsecs. Beyond this radius, however, the neutral hydrogen gas disc expands dramatically. As the radius increases exponentially, the elevation is 9.8 kiloparsecs. At the same time, the volume density and surface density of neutral hydrogen decrease exponentially. There is a close correlation between surface density, bulk density and elevation at a distance of 35 kiloparsecs from the Galactic Center. The neutral hydrogen spiral arm structure is also clearly visible at this distance. Nevertheless, unlike other galactic disc tracers, it doesn't present the genuine sliver disc spiral arm because of density and elevation perturbations. At greater distances, the Milky Way is distributed around along dark, blocky and highly-perturbing neutral hydrogen, extending into space about 60 kiloparsecs away from the Galactic Center.

Neutral hydrogen gas has a two-phase structure. In the classical two-phase model, there are two steady states of neutral hydrogen gas at pressure equilibrium: one is the cold neutral medium (CNM, T <300 K, number density = 0.3/cm³) in the form of agglomerates, the other is the diffusely-distributed warm neutral medium (WNM, T >300 K, number density = 0.3/cm³). The characteristic cooling timescales for the two phases differ by 2 orders of magnitude (100 times).

Both gases beyond the steady state belong to transient phases. The neutral hydrogen gas within 18 kiloparsecs of the galactic disc radius, and within a few kiloparsecs of galactic geysers ejecting into the halo, is in a two-phase structure. In the dynamic interstellar medium, turbulence produces density perturbations on small scales, resulting in thermal instabilities capable of enhancing phase transitions over a wide temperature range. It is observed that 50% of the gas is in the instability section (300–5000 K). The cold neutral hydrogen structure also absorbs continuous radiation against a hot background. HI Self-Absorption (HISA), as well as HI Narrow Self-Absorption (HINSA) of hydrogen, have been observed from a few parsecs to several astronomical units in the disc. HINSA outlines the atomic abundance profile in molecular clouds, and helps obtain basic parameters of star formation, such as timescale and cosmic-ray ionization, to understand the physical and chemical processes of star formation (Fig. 12).

Outside the gas layer in the galactic disc, observation shows there are many cold neutral hydrogen clouds in high-velocity clouds (HVCs) at a distance of dozens of parsecs. Deviating in velocity from the disc radiation (|Vlsr| >100 km/s), these HVCs are widely distributed from the galactic disc to high latitudes, accounting for about 10% of the total gas mass. High-resolution observation of HVCs can reveal the physical environment, such as temperature and X-ray ionization, beyond the disc and in the halo region. Statistical analysis of HVCs, along with confirmation of distance and metallic abundance (optical and infrared observation of absorption lines in the background stellar spectrum of HVCs) can help us distinguish whether the source of HVCs is consistent with the galactic geyser model of accelerated ascent of galactic disc shockwaves, and gravitational fallback, or it originates from galactic absorption of early extragalactic proto-gas, related to the amount of dark matter in the Milky Way. These unknowns will be gradually revealed through the exploration conducted by FAST, the high-sensitivity radio telescope.

With the combination of angular resolution (3′4″) and sensitivity (80 mK), the Galactic Arecibo L-Band Feed Array (GALFA) HI Survey can track objects from declination −1°20′ to 38°02′ in the northern sky. It has achieved much in the structure of the inner shell layer of the galactic disc, detection of HVCs in the galactic halo,

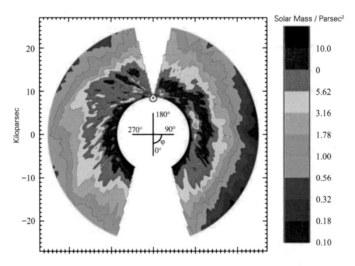

Fig. 12 The surface density distribution of neutral hydrogen in the Galaxy. The gas disc distribution in the heliosphere cannot be determined due to distance ambiguity

cold self-absorption lines of nearby gases, and compact hydrogen clouds. Compared with Arecibo, FAST has a wider field of view, observing more complete star-forming regions, such as the Orion Molecular Cloud Complex. With its large aperture and high sensitivity, FAST can observe the fainter distribution of neutral hydrogen gas at middle and high galactic latitudes. The study and statistics of high-galactic-latitude cryogenic hydrogen gas will not only refine the entire framework of the interstellar medium steady state, but also trace the structure outside the galactic disc, delving into the overall evolution of the interstellar medium.

> **Knowledge Link**
> **Trace**
>   It refers to show and record tracks and traces of the movement, evolution and development of particular substance. Tracers are capable of depicting contours, properties and motion behavior of traced objects.

1.3 Fast Radio Bursts

A new burst radio pulse phenomenon, Fast Radio Bursts (FRB), has been discovered in recent years, and relevant researches are flourishing.

While FAST has embarked on pulsar and spectroscopic observation, it has also seen FRB-related search and monitoring as important new goals, hoping to contribute to this emerging field (Fig. 13).

The first FRB was accidentally discovered by Australia's 64-m Parkes radio telescope while performing the pulsar survey of the SMC. The first fast radio burst observed, FRB010724, exhibited only a brief 5-ms pulse, but its flux was tens of janskys, quite bright by the standards of radio astronomy, and it was even enough to saturate the sensitive receivers of the Parkes radio telescope. Since the dispersion of this outburst was as high as 375 parsecs/m^3, well above typical values of the SMC, so probably it just happened to appear in the direction of the SMC, but actually originated from deeper space. Considering its brightness and duration, the FRB behavioral characteristics are not consistent with any known physical process, thus supposed to be an entirely novel phenomenon (Fig. 14).

As of May, 2017, 21 FRBs have been detected. The vast majority of these events, including the first FRB, were discovered by the Parkes radio telescope. Phenomenologically, each radio burst is also similar to the prototype, with bright pulses lasting only milliseconds. Most radio bursts are single-pulse events, with only a few samples showing a double-pulse structure; the dispersion varies from hundreds to thousands of parsecs/m^3, corresponding to cosmological distances. Consequently, based on the high dispersion of FRBs, the current prevailing theory suggests that such phenomena are likely to originate in extragalactic galaxies. The short pulse duration timescale,

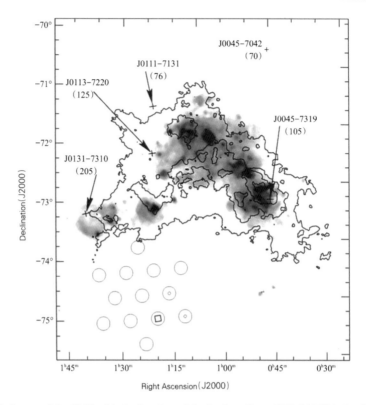

Fig. 13 Image of the SMC with the location of the Lorimer Burst FRB 010,724, the first FRB detected. The names of the individual pulsars are indicated, along with their dispersion (data in square brackets under the name of each pulsar). The 13 circles in the lower left represent the positions of the 13-beam receiver on the 64-m Parkes radio telescope during the eruption detection, with the one marked with a box corresponding to the strongest signal and the two marked with smaller circles also receiving the corresponding signals. *Credit* Lorimer et al. (2007)

in turn, indicates that the source region of the outburst is narrow and should not be longer than the distance light travels in a few milliseconds. They originate from compact stars or localized processes of ordinary stars. Given the limited field of view of radio telescopes and the fact that only the right pulses can be received at the right place at the right time, it is estimated that FRBs can occur at a frequency of thousands to tens of thousands of times per day.

Most currently known FRBs are one-time events, with the exception of FRB 121,102. The recurrence burst has been repeated more than a hundred times in recent years, each time with different photometric and radiometric bands, but with fairly consistent dispersion. Since its partial recurrence was observed in real time by the U.S. Very Large Array (VLA) and the European VLBI Network (EVN), this source is the only FRB that has been precisely and reliably located (in a dwarf galaxy at a redshift of 0.193) so far. Because quite a few samples were mined from archived observation after the outbreak, with resolution limitations of single-antenna radio

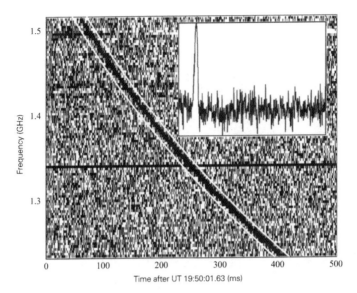

Fig. 14 Pulse profile (upper right) and frequency-dispersion plot of FRB010724, the first FRB observed. The FRB low-frequency pulses arrive later than high-frequency pulses, which is generally thought to be caused by the spread of signals in the intergalactic medium. *Credit* Lorimer et al. (2007)

antennas (and the UTMOST interferometer that detected three instances), other FRBs haven't been detected with high resolution in time to determine their origin, and their host galaxies are highly controversial.

Where do the short, bright pulses of FRBs come from? Given its extragalactic distance, short duration, and phenomenal outburst rate, coupled with the fact that the FRB 121,102 host galaxy has been determined, the most prevailing theory believes that these phenomena stem from processes related with compact star formation beyond the Milky Way. Existing models range from catastrophic events in compact stars such as double neutron star mergers, double white dwarf mergers, and the collapse of extremely massive neutron stars into black holes, to less drastic mechanisms such as magnetar giant flares (the surface shell layers of neutron stars in strong magnetic fields burst and reorganize under the influence of magnetic fields), the giant pulses of bright pulsars beyond rivers, the motion of planetary systems or small objects around neutron stars, processes relevant to young supernova relics, and intermittent brightening of low-luminosity active galaxies, and even to strange processes such as the evaporation of proto-black holes and cosmic string discharges. It is difficult to determine which model is correct because there are few samples and many uncertainties. Even fast radio bursts may have a dual origin. Recurrence bursts like FRB 121,102 may originate from repeatable events such as giant pulses or small object impacts, while other one-time outbreaks may originate from catastrophic processes (Fig. 15).

Fig. 15 Visible-light image of the host galaxy of the source of the fast radio bursts, called FRB 121,102. *Credit* Gemini Observatory/AURA/NSF/NRC

In order to make a breakthrough in the FRB study, increasing the sample size should be given priority. It will answer many knotty questions, such as whether the distribution of FRBs in the sky is really as isotropic as the cosmological distance origin theory claims? If more outbursts could be located with high precision by interferometers and certified by host galaxies, we will undoubtedly have a better understanding of the outbreak origin.

Of course, the follow-up observation of known radio bursts with greater sensitivity would also answer some unsettled questions, such as if many one-time FRBs are truly gone forever, or whether there are many recurrences undetected before because of weak luminosity. After all, FRB 121,102 was discovered by the Arecibo Observatory's 305-m radio telescope, which is more sensitive than Parkes, and a significant portion of its recurrences were very faint and not sufficiently detectable by Parkes (Fig. 16).

In the FRB research, the FAST's greatest advantage is not positioning, because even a 300-m effective aperture has a resolution of less than 30 arcmin. Compared with the Parkes telescope—the lead detector with 10 arcmin location accuracy, FAST is still not good enough to pinpoint the host galaxy.

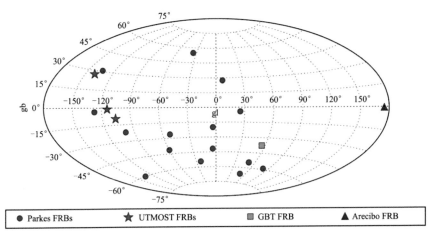

Fig. 16 The distribution of known FRBs in the galactic coordinate system. Dots mark the positions of the FRBs detected at the Parkes telescope, the triangle represents FRB 121,102 detected at the Arecibo telescope and the square represents FRB 110,523 discovered at the GBT. Stars mark the positions of the UTMOST FRBs. Due to the limited number of known FRBs and the fact that they are mostly detected by telescopes in the Southern Hemisphere, it is difficult to determine the all-weather distribution pattern of this phenomenon based on a small number of samples. *Credit* Caleb et al. (2017)

It is estimated that FAST can only detect several FRBs in every 1000 h of observation. Although it's not an unfavorable expectation, and more new FRBs will be discovered to help us determine their all-day distribution pattern, the detection rate fails to compete with large field-of-view radio arrays like the Canadian Hydrogen Intensity Mapping Experiment (CHIME), China's Tianlai Array, and the future Square Kilometer Array (SKA). The FAST's high sensitivity, its unique strength, ensures that every FRB detected can be followed up sufficiently. Furthermore, its large observable sky area can assist Parkes or other telescopes located in the Southern Hemisphere with detection. This will hopefully resolve the question of whether FRBs can recur over time. Of course, if FAST participates in the coordinated VLBI, it will become the leader of VLBI network in East Asia by virtue of its large receiving area. Hopefully, it enables the precise localization of FRBs while balancing sensitivity, and ultimately determines the origin of these short but bright bursts.

1.4 Obtaining Ultrafine Structures of Celestial Bodies

In astronomy, angular resolution means the minimum distance between two distinguishable objects. Resolution angle $\theta = \frac{\lambda}{D}$, the radio of wavelength to aperture. Radio astronomical telescopes operate at wavelengths millions of times greater than those of optical telescopes. To have their resolution comparable to that of optical telescopes, the dishes' diameter needs to be hundreds of kilometers or the size is as big as the Earth. Besides, their error has to be limited to 1 mm or less, and this is unavailable with existing technologies. Radio astronomers have found ways to improve the resolution without increasing antenna apertures—Radio interferometry, which eventually developed into VLBI. The two antennas incorporating VLBI are able to cross continents and oceans. Their angular resolution is $\theta = \frac{\lambda}{B}$, Baseline (B) can be as long as the diameter of the Earth, and it will be longer if antennas are sent into space. The resolution of the modern global VLBI network is finer than a milliarc-second and at least 1000 times better than the resolution of all other astronomical bands.

If you share a cake for 12 people, each will have a circular sector with the central angle of 30°; if everyone on Earth were to divide the cake, each would still get one with the central angle much larger than the resolution angle of VLBI. Whether a radio telescope can join the VLBI club, and what role it plays, partly reflect its display resolution. The major VLBI networks in the world are the European network, the United States network and the Asia–Pacific network. Their main unit antenna apertures are 20–40 m, and the largest unit antenna aperture is 100 m. If FAST joins the VLBI, with its large receiving area and geographical advantage, it will become the leader of international VLBI, marking the China's leading position in international cooperation in this field.

The development of VLBI techniques and image reconstruction algorithms has changed the face of astronomy in many areas over the past 30 years: Imaging of distant quasars and proto-stellar galaxies; discovering the fine structure of active galactic nuclei (see Fig. 17); directly revealing the transport of matter and energy; and providing observational facts for modelling the central engines of early stellar systems.

With the FAST participation in the intercontinental VLBI network, the baseline detection sensitivity can be increased by 5 times. The resolution of VLBI networks is not only related to the length of the longest baseline, but also to its weight. FAST is at the edge of all international networks, and the high sensitivity gives a high weight to the FAST-related baselines, thus enabling higher resolution to the VLBI networks in which it participates. Its leading role is more striking when it collaborates with space-orbiting radio telescopes. If the ground network of the Japanese 8-m-diameter VSOP space telescope were to include FAST, the 8-m telescope would be equivalent to 100 m, and the number of targets that could be mapped would increase by 1000 times. If FAST replaces the Arecibo telescope in the High Sensitivity Array (HSA) observation consisting of the U.S. Very Long Baseline Array (VLBI), the Very Large Array (VLA), the Green Bank Telescope (GBT), and the 100-m radio telescope at Bonn, Germany, the HSA's sensitivity will be increased from 5.5 to 3.1 µJy, and its FoV will be enlarged. Coupled with the FAST's unique geographic location and much longer trackable observation time than the Arecibo, the UV coverage of VLBI networks can be greatly enhanced, thus improving image quality.

Except for a small number of phase-reference modes, the detection integral time of VLBI is limited to minutes or seconds because of signal coherence, thus it can image very few targets compared with the wired interferometer array. Of the 3035 celestial objects in the most complete U.S. VLBI source list, only some are imaged,

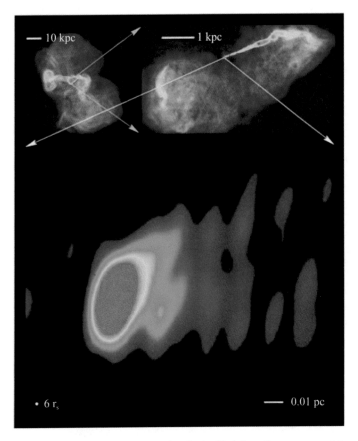

Fig. 17 Radio images of the galaxy M87, showing the double-lobe radio structure and the one-sided jet

of which about 300 have experienced the multi-epoch, multi-frequency monitoring studies, and more than 50 have been accurately imaged in polarized magnetic fields. Only 30% of the sources with fluxes greater than 1 Jy in the NRAO-MPIfR 5 GHz radio source table have VLBI imaging observation, and intragalactic sources are even fewer. With the FAST participation, the detectable targets will be increased by at least two orders of magnitude, providing complete statistical samples of dense sources to more reliably test theories and models of the Active Galactic Nuclei (AGNs) core engine and discover amazing phenomena in deep space. For instance, FAST will greatly advance the mapping observation of faint, high-redshift (z >3) AGNs with high spatial resolution, eventually significantly expanding the statistical sample capacity of the proper motion-redshift relationship in cosmology. As another example, in observational studies of apparent superluminal sources, the brightness decreases rapidly when radiating masses in jet streams move away from the center. Currently, the sensitivity of telescopes is only sufficient enough to conduct tracking observation for several years, limiting multiple superluminal studies, and FAST will radically change the situation. The polarized radiation from the radio source core accounts for only 1% of the total intensity emission, and accurate imaging of magnetic fields in gravitational cores of distant galaxies will indeed lead us to the uncharted territory of physics.

If VLBI includes FAST, the 10-m space telescope and the 100-m ground telescope, it will be 50–100 times more sensitive than the existing equipment. It's possible that VLBI has access to a few fine images of thermal spectral sources by right of its resolution better than 0.1 solar-terrestrial distance. This makes it easier to study the formation and evolution of stellar-like objects, even to directly image Jupiter-like planets in nearby stellar systems.

1.5   Exploring the Origin of Life in Space—Interstellar Molecules

Interstellar molecular spectral lines were one of the four major astronomical discoveries of the 1960s. In 1952, Miller conducted laboratory experiments on the origin of life on Earth. By using hydrogen, ammonia, methane and water to simulate the primordial atmosphere and oceans and injecting energy through electric shocks, he produced a variety of organic molecules associated with life processes. Radio astronomy suggests that the appearance of complex molecules in the pre-life period may not have to start from the very beginning, and in the early 1960s, thanks to the development of millimeter-wave astronomy, spectral lines of molecules resulting from jumps in different rotational energy levels were observed in the interstellar medium. These molecules included the basic chemical elements which are able to write protein formulas, such as C (carbon), H (hydrogen), N (nitrogen), and O (oxygen). Townes, the founder of molecular astronomy, was awarded the 1964 Nobel Prize in Physics.

Knowledge Link

Masers

They are naturally occurring sources of stimulated spectral line emission caused by the population energy level layout inversion. This is an extreme deviation from thermodynamic equilibrium, thus called non-thermal maser spectral lines. The molecules emitting maser spectral lines are called maser molecules, and those celestial objects that produce maser radiation are called maser sources.

Megamasers

They are celestial objects associated with active galactic nucleus. Their radiation processes, motions, special physical environments, and relations to central objects have become frontier research programs.

Late-type stars
They are type K and M stars in the stellar spectral sequence.

By 2005, 129 interstellar molecules had been identified, including eight maser molecules with more than 50 non-thermal maser spectral lines. Thousands of maser sources have been discovered in the Milky Way, and 106 hydroxyl (OH) megamaser sources and 64 water megamaser sources have been found in extragalactic galaxies.

Interstellar molecules are found in a wide range of astronomical environments. About 20% of the molecular spectral lines are in the centimeter and decimeter bands. Since stars derive from molecular clouds, and different molecules and lines can trace various physical conditions, which in many cases can only be observed by molecular lines, the observation of molecular lines is essential for studying the stars' formation and evolution.

As a prerequisite for the star formation, the molecular cloud formation marks the condensation of a large amount of gas. The molecular gas detection in extragalactic galaxies also plays an important role in determining the morphology and evolution of galaxies, especially in high-redshift galaxies and proto-galaxy candidates, since the formation of certain molecular spectral line sources (e.g., maser sources) requires specific conditions related to the types of galaxies and the evolution stage of galaxy nuclear activity. Narrow molecular spectral lines facilitate accurate determination of galaxy redshifts, so the FAST centimeter-band and decimeter-band molecular spectral line studies, the Molecular outward flow and neutral hydrogen linkage measurements, etc., will greatly promote the development of molecular spectroscopy research in China.

Maser sources and extragalactic maser sources have strong radiation and small spatial scale. The minimum scale of individual maser spots is close to one astronomical unit. These spots are found in star-forming regions and in the vicinity of late-type stars, and their VLBI observation is the best tool for studying the physical and dynamical conditions in the small-scale environment of the Milky Way and nearby galaxies. The observation of masers in the Galaxy as well as extragalactic galaxies makes huge contributions to astrophysical research such as dynamics of molecular clouds, star formation, interstellar magnetic fields, determination of the galactic scale and distance to nearby galaxies, and black hole certification. Current maser research will turn to extragalactic megamasers and extragalactic megamaser bursts.

FAST is designed to contain the spectra of 12 molecules, including hydroxyl (OH) and methanol (CH_3OH), within its operating bandwidth. The FAST's high sensitivity enables the extensive search for OH and CH_3OH megamasers in extremely bright infrared galaxies, high-redshift galaxies, active galaxies, and quasars.

While the Arecibo telescope is a pioneer in the megamaser detection, FAST can observe more hydroxyl megamaser sources, and further study the relationship between megamasers and types of galaxies, nuclear activities, and relativistic outflows in galactic nuclei.

We have now obtained evidence for the existence of a black hole in galaxy NGC4258 through megamaser observation. Large samples of hydroxyl megamasers will make it possible to obtain more evidence for the existence of black holes. Using the Arecibo telescope, astronomers have detected the brightest hydroxyl megamaser at the redshift of 0.6. However, FAST can detect it at higher redshifts ($z \sim 1$), making the cosmological study of hydroxyl megamasers possible. At present, the photometric functions of hydroxyl megamasers are poorly determined, and physical mechanisms of these masers remain unknown. The sky survey of hydroxyl megamasers with FAST in the multi-beam mode will help us better know their photometric functions, providing essential information about their origin. Megamasers are the brightest radio point sources in the Galaxy, nearly 100 times stronger than the nearby hydroxyl masers. Methanol masers are necessary to trace the star and planet formation and study accretion disks. The international community has failed to detect methanol masers so far. Given that FAST has a larger field of view than the Arecibo telescope, it's possible that we are going to find the first extragalactic methanol megamaser with the FAST's extremely high sensitivity. It could detect gigamaser galaxites at high redshifts, thus assisting scientists to explore properties of the cosmic evolution at early stages.

1.6   Are We Alone?—Exploring Extra-Terrestrial Civilization

"Who are we? Where do we come from? Are we alone?" We wonder if there are any other civilized societies beyond Earth. The philosopher Russell once said, "There are two answers to this question, yes and no, and both will surprise us equally." The risk of the "Search for Extra-Terrestrial Intelligence (SETI)" subject speaks for itself. However, once it succeeds, all mankind's scientific achievements will be outshone. That's why the scientific community has never stopped exploring, and the governments and citizens of developed countries have never stopped investing in SETI.

Life has been found alive in "extreme life-environments" unimaginable to humans: hot springs under the sea at hundreds of degrees Celsius, the sky at an altitude of hundreds of thousands of meters, and rock formations several kilometers underground.

Life's resilience is beyond human's imagination. Therefore, when we consider extra-terrestrial life, pleasant environment is not our only concern. The Galileo spacecraft reached Jupiter in 1995 and sent back images showing that there was much more water under the icy crust of Europa than on Earth; In 2004, the European "Mars Express" took pictures of water–ice at the south pole of Mars. NASA's Spirit rover (MER-A) and Opportunity rover (MER-B) landed on the surface of Mars and have been working ever since, revealing the planet's previous wet periods through extensive geotechnical sampling; In 2005, Cassini-Huygens successfully orbited Saturn, and Huygens landed on Titan, confirming the existence of water–ice and hydrocarbons. The Sun is one of billions of stars in the Galaxy, since 1986, astronomers have discovered thousands of extra-solar planets by virtue of high-precision line-of-sight (LOS) velocimeters. With advances in science—Earth's extreme life environment, extra-terrestrial water, and extrasolar planetary systems, SETI has been paid more attention.

If our civilization followed the philosophical "law of mediocrity", according to the Dracula Green Bank formula, we should have many civilized neighbors. How come we never hear from them? How can we find them? So far, no imprints of life have been found on other planets in the solar system, and it can almost be concluded that complex forms of life do not exist there. The law of the speed limit of light, the vast distances between stars, and the incredible energy consumption make interstellar travel long and unreachable, thus mainstream science doesn't consider interstellar travel to be feasible. The only viable way for us to communicate with extra-terrestrial civilization is to look for "artificial" radio signals from beyond the Earth.

Knowledge Link

Drake (Green Bank) Equation

It was written in 1961 by Frank Drake, an American astrophysicist, to estimate the number of active, communicative extra-terrestrial civilizations (defined as any civilization that has mastered radio astronomy.) in the Milky Way. We could communicate with any civilization which masters radio astronomy with existing technologies on Earth.

Non-thermal galactic background noise, quantum noise, and cosmic microwave background noise are three ubiquitous sources of noise in our galaxy. Faced with the same electrical noise spectra, engineers of the extra-terrestrial civilizations and our scientists would probably share the same frequency windows (see Fig. 18).

SETI experts believe that humans should concentrate their search in the 1–3 GHz frequency range, especially between the 21 cm neutral hydrogen (HI) line and 18 cm hydroxyl (OH) line. H combines with OH to form water, so this narrow band is also known as the "water tunnel". Water is fundamental to life on Earth, and extra-terrestrial "aquariums" may also naturally seek their own kind through the "water tunnel".

The SETI study has attracted great interest since the ground-breaking article entitled "Searching for Interstellar Communications", written by Giuseppe Cocconi and Philip Morrison, was published in *Nature* in 1959. Of the many programs currently available, the Phoenix Program is one of the most comprehensive SETI programs. It was launched in 1994, applying the world's largest antenna to systematically search for radio signals from about 1,000 nearby sun-like stars.

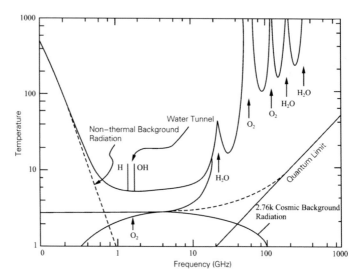

Fig. 18 Possible windows for interplanetary radio communication

In 2006, the ATA Telescope Array, privately funded by U.S. enterprises, became partially operational. Dedicated to SETI science and interstellar communication detection, it consists of 350 antennas, with a scale of 300 m × 200 m.

1.7 The FAST Application Objects

FAST will extend China's space measurement and control capabilities from geosynchronous orbit to the outer edge of the solar system; increase the downlink rate of deep space communications data by tens of times; increase the measurement accuracy of the ToA of pulsars from 120 to 30 ns, making it the most accurate pulsar timing array (PTA) in the world; make pulsar clocks for the prospective research—autonomous navigation; conduct high-resolution microwave patrols to diagnose and identify weak space signals with 1 Hz resolution in the service of national security as a passive strategic radar; serve as an incoherent scattering radar (ISR) receiving system for the "Meridian Project", providing high resolution and observation efficiency; and track the coronal mass ejection (CME) detection for space weather forecasting.

The FAST construction involves many high-tech fields, such as antenna manufacturing, high-precision positioning and addressing, high-quality radio receivers, sensor networks and intelligent information processing, ultra-wideband information transmission, massive data storage and processing, etc. The key technological achievements can be applied to many related fields, such as large-scale structural engineering, kilometer-range high-precision dynamic measurement, the development of large industrial robots, and multi-beam radar devices. Moreover, its construction experience will greatly promote the information-based, cutting-edge and eco-friendly development of manufacturing technologies in China.

2 Innovative Concepts of FAST

The FAST's active reflector converging electromagnetic waves serves as the lens of the eye, the feed receiving waves can be seen as the retina, and the actuator controlling the reflector and the cable-driving system pulling the feed cabin act like muscles that control the eyes. Consequently, FAST is vividly called the "China Sky Eye", and the orbit that houses it is a Karst depression named "Dawodang".

Finding a suitable orbit for the "China Sky Eye" requires a large, round karst crater with engineering and hydrogeological conditions that meet its operational requirements. This is not an easy task. Various innovative site selection methods and technologies have played an important role in this work.

With the support of remote sensing and antenna data, scientists have found a proper orbit for the "China Sky Eye" through innovative scoring mechanisms and fieldwork.

Laid in the karst depression, the active reflector is the "crystalline lens" of the Eye. The active control enables a part of the 500-m aperture active reflecting spherical cap will form a 300-m aperture instantaneous parabolic surface in the observation direction, converging the electromagnetic waves on the focal point. In this way, the Eye can see the celestial objects clearly without scattering the light.

The crux of designing the active reflector lies in the focal ratio of ~0.4611. The maximum difference between the parabolic surface and the central sphere is only 0.67 m in the 300 m aperture illuminated area. These small displacements can be easily adjusted by electromechanical control to deform a portion of the central sphere into a parabola, which makes possible the broadband observation by applying receiving technologies of traditional telescopes (Figs. 19, 20 and 21).

To make the Eye rotate flexibly, it is necessary to control the flexible movement of the feed cabin on the focal plane. On the 500 m huge spatial scale, it's impossible to form a rigid connection between the receiver at the focal point and the reflecting surface. Meanwhile, the pointing and tracking platform solution for the Arecibo telescope would be more than 10,000 tons, which means that the solution is unfeasible. As a result, over the years, scientists have developed sophisticated cable-dragging technology of optical, mechanical and electronic integration, which achieves high-precision pointing and tracking by using six cables to drag the receiver's feed cabin to the focal point, with a finely tuned robot to counteract the vibrations of cables.

The feed cabin is equipped with a multi-band, multi-beam feeder and a receiver system covering a frequency of 70 MHz–3 GHz. It is the "retina" of the "China Sky Eye" (Figs. 22, 23 and 24).

Fig. 19 The original appearance of the Dawodang depression, the FAST site

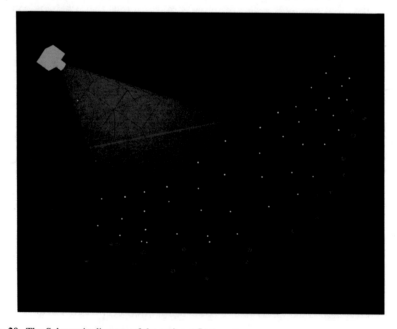

Fig. 20 The Schematic diagram of the active reflector

Fig. 21 The side view of the active reflector

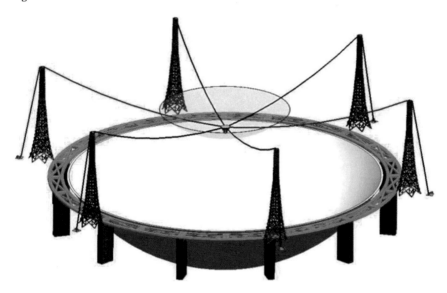

Fig. 22 The cable-driving system

Fig. 23 The design of the light-weight feed cabin

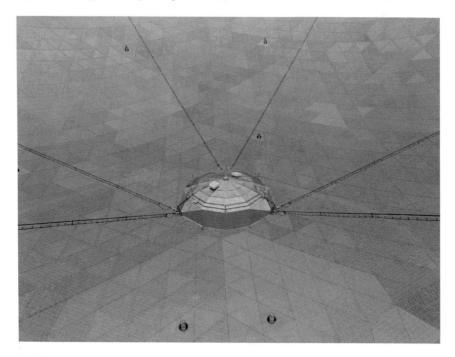

Fig. 24 The light-weight cable-driving feed cabin

3   System Composition of the FAST

3.1   Building the Orbit for the Eye—The Site Construction

Scientists carried out a detailed survey of the topography, engineering geology and hydrogeological environment of the selected area. Although the Dawodang's envelope shape is very similar to the reflecting spherical cap, the natural depression has to be further shaped, thus more than 1 million m^3 of rock and soil have been excavated. Although the depression is well-drained, vertical and horizontal drainage channels need to be dredged and constructed to ensure that the electromechanical equipment at the base of the telescope is not at risk of flooding in a 50-year rainstorm.

3.2   The Eye's "Crystalline Lens"—A Movable Reflecting Surface

The back frame of the FAST's active reflector is a steel cable net structure, with thousands of triangular cable net configurations with 11 m side lengths formed by nearly 10,000 steel cables, and the triangular vertices are connected by about 2300 nodes.

A total of 4450 reflector panels were laid on top of the cable, including 4300 triangular reflector panels and 150 quadrilateral reflector panels. Based on astronomical coordinates and real-time measurement of the reflecting surface shape, the observation controls the reel mechanism on the ground. Winch nodes enable the mechanism to be displaced to complete the active deformation of the reflecting surface from spherical shape to parabolic shape.

3.3   The Feed Support

The feed support system consists of six support towers which are over 100 m high around the depression, and a flexible supporting system made of steel cables on the kilometer scale as well as its guiding ropes and reel structures. It can adjust the first space position of the feed cabin, which serves as the "eyeball" of the "China Sky Eye". The FAST feed cabin is about 13 m in diameter, equipped with an AB axis steering mechanism and parallel robots for secondary adjustments to compensate for the vibration of primary cables and to achieve 10 mm spatial positioning accuracy of the feed. Meanwhile, it's necessary to build power and signal channels from the ground to the feed cabin, as well as safety and health monitoring systems, including lightning protection, cable stress monitoring, and emergency prevention and equipment.

3.4   High-Precision Measurement and Control

The main structures are in motion when FAST is operational. There is no rigid connection between the main reflector and the feed. Long range, high sampling rate, and high precision measurement and control are key constraints for building a successful telescope. The system involves the construction of reference networks with millimeter precision, the installation of GPS and laser trackers, and the scanning of 1000 control points within the illuminated area. Therefore, a large-scale fieldbus needs to be built to achieve active deformation control of reflective surfaces, and advanced dynamic decoupling control techniques are worth developing in order to realize the receiver's spatial positioning.

3.5   Feeds and Receivers

According to the FAST's scientific objectives, the operating frequency covers 70 MHz–3 GHz, including seven receivers and terminals (the core is the 19-beam receiver in the L-Band), low-noise refrigeration amplifiers, broadband digital IF transmission equipment, high-stability and high-precision time–frequency standard equipment. It's also necessary to develop a receiver's operating condition monitoring and fault diagnosis system with multi-purpose digital astronomy terminals.

The feed receives the electromagnetic waves converged by the primary reflecting surface, the low-noise amplifier amplifies signals to the appropriate intensity, and the radio frequency (RF) filter selects desired bands for observation. RF amplifiers, mixers and IF filters further process signals, and optical fibers transmit intermediate frequency (IF) signals to data processing terminals in the ground observation room.

3.6   Observatory Construction

The observation base is essential to support the operation, observation and maintenance of the telescope. Depending on functional requirements, the base consists of multiple-use buildings, maintenance workshops and other buildings scattered around the base and the reflector (Fig. 25).

4   Operating Principle of the FAST

The basic principle of a classical radio telescope is similar to that of an optical reflecting telescope. Reflected by a precise mirror, electromagnetic waves arrive in phase at a common focal point. Using rotating parabolas as mirrors is easy to

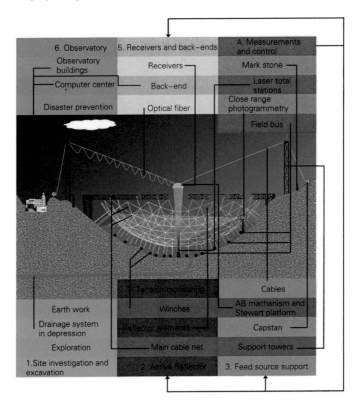

Fig. 25 The FAST construction specification

realize the in-phase focusing, and most radio telescope antennas are parabolic. For the observation of meter waves or long decimeter waves, a metal mesh can be used as a mirror. In contrast, for the observation of centimeter and millimeter waves, a smooth, precise metal plates (or coating film) is used as a mirror. Antennas or antenna arrays send the collected astronomical waves through feed lines to receivers. Receivers are extremely sensitive and stable, and they detect the weak astronomical waves after high amplification, converting the high-frequency signals into low-frequency ones for recording. By analyzing these curves, astronomers get all kinds of cosmic information sent by celestial bodies.

The basic metrics that characterize the performance of radio telescopes are spatial resolution and sensitivity. The former refers to the ability to distinguish between two radio sources that are close to each other. The higher the resolution, the closer the two sources can be separated. The latter means the "minimum measurable" energy value of a radio telescope. The lower the value, the greater the sensitivity. Radio telescopes are usually required to have high spatial resolution and high sensitivity. Common methods to improve sensitivity include reducing the inherent noise of receivers, increasing the receiving area of antennas, and extending the observation integration time.

Because of the earth's atmospheric shielding, only radio waves from celestial bodies at wavelengths of about 1 mm–30 m can reach the ground. To date, most radio astronomy research has been conducted in this band.

Conventional parabolic telescopes converge the incident planar electromagnetic waves into a focal point, while the spherical reflective surfaces converge into a focal line, limiting the application of wideband polarization receivers. To tackle the problem, the FAST's spherical crown reflecting surface forms a 300-m aperture instantaneous parabolic surface in the direction of radio sources, making the receiver in the feed cabin to be placed in the same focal point as conventional parabolic antennas.

The FAST's spherical reflector will be divided into 4450 units controlled by actuators to achieve active deformation of the reflector. Due to the large spatial span, it is difficult to establish a rigid connection with high precision between the receiver and the reflecting surface.

FAST adopts a mechanism-electronics-optics integrated, cable-supported and lightweight feed platform with secondary adjustments in the feed cabin to position receivers in space. At the same time, to complete the active deformation of the reflecting surface and accurate pointing and tracking, it adopts the measurement characterized by large range, high precision and high sampling rate as well as corresponding control technologies.

The feed cabin is equipped with international advanced high-quality multi-beam receivers, which are used to collect the cosmic radio waves converged by the reflecting surface and transmit them to the terminal through broadband optical fibers to analyze the astrophysical information obtained.

FAST is 2.25 times more sensitive than the Arecibo telescope. Moreover, the operating limit of the Arecibo is within 20° of the zenith, which restricts its observation field, especially the networking observation capability.

5 Advanced Technologies of the FAST

See Fig. 26.

5.1 High Sensitivity

For creatures, it's not necessarily the case that the bigger the eye, the better the vision, but for telescopes, particularly for radio telescopes, that is the case.

The primary indicator of a radio telescope is the receiving area, which represents its ability to observe faint objects. FAST is currently the most sensitive radio telescope in the world. Before that, the Effelsberg 100-m Radio Telescope in Germany and the 100-m Green Bank Telescope in the United States were the world's two largest fully steerable single-aperture telescopes. However, FAST is nearly 10 times more sensitive than them.

5.2 Large Sky Area

Because of the enormous size and depth of the Eye's orbit—the Dawodang Depression, and the Eye's muscle—active reflector control system and mechanism-electronics-optics integrated feed support system, FAST is able to view objects within 40° of the zenith, two times larger than that of the Arecibo telescope. If the developing focal plane array feed technology is applied, its field of view (FoV) will be within 60° of the zenith, providing more opportunities for the telescope.

5.3 Broadband Coverage and Multiple Beams

FAST operates at frequencies ranging from 70 MHz to 3 GHz, which is closely related to the precision of the "crystalline lens" reflecting surface and the "retina" feed. The design is relevant to the telescope's scientific objectives, and recent advances in radio astronomy tell us that low-frequency bands are full of opportunities to make progress. For instance, the search for cosmic dark matter and dark energy by observing neutral hydrogen at high redshifts; the detection of the first generation of luminous objects; the discovery of more pulsars; the search for quark matter and neutron star—black hole systems. The FAST frequency coverage is up to 3 GHz, enabling itself to carry out high-precision pulsar timing measurement while participating in extensive international cooperative measurement as a mainstream telescope. Its multi-beam feed will increase the efficiency of conventional radio astronomy observation by 10–20 times.

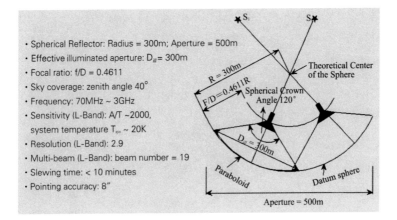

Fig. 26 Main technical specifications of FAST

Pillars of a Great Power—Intelligent Manufacturing in China

Haiyan Zhang, Lei Qian, Caihong Sun, Chengmin Zhang, Wenjing Cai,
Aiying Zhou, Chengjin Jin, Li Xiao, Dongjun Yu, Qing Zhao, Boqin Zhu,
Wenbai Zhu, Lichun Zhu, Ming Zhu, Liqiang Song, Mingchang Wu,
Baoqing Zhao, Ming Zhu, Gaofeng Pan, Hui Li, Rui Yao, Youling Yue,
Bo Zhang, Rurong Chen, Boyang Liu, Li Yang, Na Liu, Jiatong Xie,
Yan Zhu, Hongfei Liu, Zhisheng Gao, and Xiaobing Chen

FAST is designed, developed, manufactured and constructed by Chinese scientists. It took 22 years from the pilot study in 1994 to its completion in 2016. Its design and construction witnessed three major independent and endogenous innovations. So how was the "China's Sky Eye" built? What difficulties and challenges were encountered in the construction process? Let's take a closer look at the world's largest single-dish radio telescope.

The steel structure of the ring beam of FAST was successfully completed in December 31st, 2013.

© Zhejiang Education Publishing House 2021
R. Nan (ed.), *The Sky Eye*, China's Big Science Facilities,
https://doi.org/10.1007/978-981-16-3824-4_3

1   Let the Dream Set Sail

As an old Chinese saying goes, it takes 10 years to grind a sword. The same is true for the birth of FAST. It took 22 years from the proposal of the project, the formation of the concept, pre-research process, feasibility study, national project approval, to official commencement, comprehensive construction, and completion.

The International Union of Radio Science held a meeting in Kyoto, Japan in 1993 to draw the blueprint for the development of radio astronomy in the early 21st century. Astronomers from China, Australia, Canada, France, Germany, India, the Netherlands, Russia, the United Kingdom and the United States analyzed the growing trends of the overall performance of radio telescopes. To observe the neutral hydrogen at different cosmic distances, they proposed the construction of the next-generation large radio telescope (LT), a radio telescope array with a total receiving area of 1 square kilometer (LT was renamed Square Kilometer Array (SKA) in 1999). Scientists dream to take a real and good look at the primordial universe and figure out how the structure of the universe was formed and evolved to what it is today before the radio environment is completely destroyed. The large radio telescope is the only tool to help make this dream a reality, and if the boat were missed, human being would have to go to the far side of the moon to build telescopes of the same size to achieve the goal.

In this context, the former Beijing Astronomical Observatory of the Chinese Academy of Sciences proposed a Chinese solution to build an Arecibo-type LT, originally named KARST, by making full use of the karst landforms in southwest China. In February 1994, the Beijing Astronomical Observatory established the Research Group of LT Advancement. In the same year, in cooperation with the Institute of Remote Sensing Applications, Chinese Academy of Sciences, and supported by many departments of Guizhou province, the Beijing Astronomical Observatory started working on the site selection for LT in China. Since 1994, site selection experts have made high-resolution digital topographic maps for 90 depressions out of more than 400 depressions that they have visited, and, after repeated comparison and analysis, finally selected the Dawodang depression in the Jinke village, Kedu town, Pingtang county, Qiannan Prefecture of Guizhou province as the site of the telescope.

In November 1995, led by the Beijing Astronomical Observatory, the Committee for the Advancement of the Large Radio Telescope in China was established with members from more than 20 Chinese universities and research institutions, and Researcher Nan Rendong served as the Director (Figs. 1, 2 and 3).

In the process of improving the concept of KARST and discussing with the international scientific community, Chinese scientists formed a solution of Arecibo-type (i.e. spherical reflector antenna array) LT by making use of the karst landform in Guizhou province. It will be a mammoth project consisting of more than 30 antennas with a diameter of 300 m or so covering an area of hundreds of kilometers, with resolution capabilities ranging from 1 arcminute to 100 milliarcseconds, which fundamentally enhanced the imaging capability of radio astronomy community. In order

Astronomers Sign International Agreement to Plan Square Kilometre Array

Fig. 1 The proposal of the concept and preliminary design of International square kilometer array

to promote the implementation of this plan, the Chinese scientists came up with the idea of independently developing a new single-aperture giant radio telescope, the Five-hundred-meter Aperture Spherical Telescope (FAST). It took more than a decade of repeated tests to perfect the idea. During this period, a number of Chinese scientific research institutions have carried out cooperative research for 14 years on 5 key technologies, including the site survey, the active reflector, the feed supporting system of optical, mechanical and electronic integration, the high-precision measurement and control system, and the receiver, gathering the innovative concepts of many Chinese scientists from different research institutions (the concept of platform-less feed supporting system, the concept of active deformation of the reflector and the concept of moving-car feed supporting system, etc.). On July 10, 2007, FAST project was approved by the National Development and Reform Commission. On December 26, 2008, the Foundation Stone Laying Ceremony for FAST Project was held at Dawodang depression. On March 25, 2011, the construction of FAST was officially started (Figs. 4 and 5).

FAST project pooled the wisdom and contribution of more than 100 old, middle-aged, and young astronomers and related scientific and technological workers from home and abroad. It was well understood and supported by Guizhou province and its

Fig. 2 China's LT solution was put forward

Fig. 3 The expert group were looking for a site in Guizhou province

Fig. 4 International review and advisory conference on FAST

Fig. 5 The foundation stone laying ceremony for FAST project

prefectures, counties, towns and villages as well as by the relevant state science and technology competent authorities (Ministry of Science and Technology, National Development and Reform Commission, National Natural Science Foundation of China, Chinese Academy of Sciences, etc.). Tens of thousands of workers from 50 enterprises made painstaking efforts for it and the national and international astronomy community, together with 39 million people from Guizhou province paid close attention to and looked forward to it.

FAST will provide scientists with significant discovery opportunities in pulsar detection and timing (gravitational wave detection) and neutral hydrogen detection (dark energy and dark matter detection). It will also serve as the cradle for Nobel Prize in Physics, ushering in a "dialogue" between human being and extraterrestrial being and potentially bringing the former's being alone to an end.

2 Find a Home for China Sky Eye

2.1 Why Karst Depression?

The construction of the giant spherical reflector radio telescope required natural depressions, a feature that only developed in karst areas. Since 1994, the FAST project team had carried out multidisciplinary site assessment in the karst area in southern Guizhou province by remote sensing, geographic information system, Global Positioning System (GPS), field trips and computer image analysis. The team made a preliminary assessment in such aspects as the natural geography, factors controlling geomorphological development, the shape characteristics of depressions, hydrogeology, engineering geology, meteorology and radio environment; according to the envelope of the depression, the team assumed the parameters of the reflector and its location in the depression and calculated the volume of filling and digging with fitting method, and offered optimization results. Finally, they completed the preliminary comprehensive surveying of the preliminary candidate sites in engineering geology and hydrogeology. Many experts from home and abroad (including two directors of Arecibo Observatory) visited the preliminary FAST candidate sites, and expert groups from FAST project also visited Arecibo Observatory in 1999 and 2006 respectively. They all believed that the preliminary candidate depression is the unique site for large spherical radio telescope in the world.

The candidate sites should be relatively accessible so as to save unnecessary investment. The fact is that, fortunately, the candidate area is rich in water conservancy and coal resources with dozens of large and medium-sized hydropower stations in place. In addition, the province hosts a number of enterprises in electronics, construction, materials, machining, energy and transportation who may participate in the bidding for FAST construction.

After a comprehensive assessment of the preliminary candidate sites, scientists selected the final site in the Dawodang depression, Jinke Village, Kedu Town, Pingtang County, Qiannan Buyei and Miao Autonomous Prefecture in the karst landform in southern Guizhou province (Figs. 6, 7, 8, 9, 10 and 11).

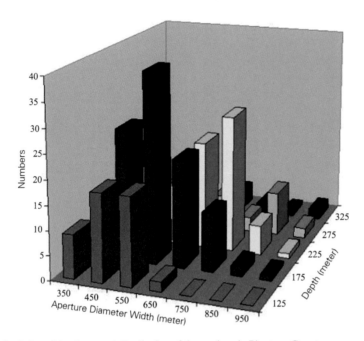

Fig. 6 Statistics of the shape and distribution of depressions in Pingtang County

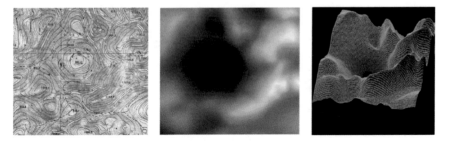

Fig. 7 The Terrain, DTM image and 3D display of Dawodang Depression

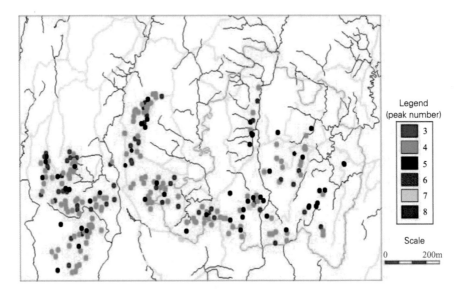

Fig. 8 Distribution map of Pingtang Karst depressions (including peak number)

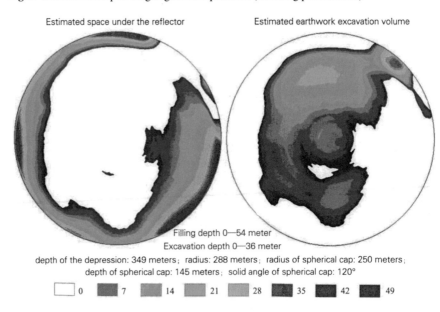

Fig. 9 Analysis of fill and excavation in Dawodang depression with a spherical diameter of 500 m and an opening angle of 120°

Fig. 10 The investigation site of FAST and the physical and mechanical properties of soils, rocks and the cores collected

2.2 Essential Quiet Radio Environment

As mentioned earlier, radio signals from the universe are so weak that they are extremely vulnerable to radio interference from human activities. Radio telescopes need a quiet radio environment. But the actual situation is that the development of the global economy is posing a serious challenge to the radio environment of radio astronomy.

As one of the radio services defined by the International Telecommunication Union (ITU), radio astronomy service is to receive radio signals from the universe and combine astronomy with radio. In 1933, Karl Jansky, an engineer from Bell Telephone Laboratories, first detected and confirmed radio radiation from the galactic center of the Milky Way at a frequency of 20 MHz, opening a new chapter for astronomy that was traditionally based on optical observations. Since its birth, radio astronomy has reaped fruitful results in astronomy research and national defense application.

Fig. 11 Image of Dawodang depression captured by QuickBird satellite

Compared with other services, radio astronomy has the following features:

(1) High sensitivity. Objects observed by radio astronomy often lie several billions or even tens of billions of light years away in the deep universe. Radio signals from celestial bodies are extremely weak, requiring the radio astronomical receiver to be highly sensitive. In order to obtain high sensitivity, the bandwidth of very-long-baseline of the continuous spectrum needs to be hundreds of MHz.

(2) Radio astronomy cannot select frequencies arbitrarily. The spectrum of celestial radiation can be divided into two categories: continuous spectrum and spectral line. The former is mainly produced by the thermal and non-thermal radiation and is generally wide; The latter is mainly produced by the energy level transition of the atom or molecule and is generally narrow. Due to the different radiation mechanisms, the scientific goals corresponding to each frequency band are different, and the scientific output and significance are also different.

(3) Passivity. Given that radio astronomy does not emit and it is passive, so it will not interfere with other services. The earth exploration-satellite service and space research operations that do not use man-made emitters are also passive operations. But wireless communication systems in our daily life such as radio broadcasting, mobile phones, cluster intercom as well as other services using man-made emitters like radar, radio beacons are called active services.

In order to protect the radio environment of the radio astronomy observatory, the first step to reduce interference is to designate the radio quiet zone (RQZ), which is the common practice of most radio observatories in the world to ensure the normal radio astronomical observation. Pursuant to the Radio Regulations of ITU, the operation of radio observatory stations that do not conform to the Table of Frequency Allocations is allowed under a given condition.

The radio quiet zone doesn't mean that radio is prohibited in the zone. For example, it is required by the Radio Regulations that all emissions are prohibited in the bands 1400–1427 MHz, but there is still interference from the out-of-band emission. What the radio quiet zone protects against is harmful radio interference. Measures taken to mitigate radio interference will affect the definition and scope of the radio quiet zone.

The international radio quiet zone includes the far side of the Moon and the Lagrangian Point L2. While the national radio quiet zone is established by a competent authority to regulate ground operations and is independent of international radio regulations, with little or no impact on satellite operations. The radio quiet zone usually consists of two different protection zones.

(1)   Radio limited zone. This area, usually designated by a country or a local government, is generally a circular or elliptical area centered on radio telescopes, controlling radio interference from power feed and electronic equipment and covering an area of a few kilometers to tens of kilometers, within which the heavy industry will be limited, no more new radio emission service will be provided, cars and tractors will not be allowed to pass and wideband radio equipment that is hard to control such as microwave ovens and medical equipment are prohibited.

(2)   Radio coordination zone. This is an area where no newly established emitter can have more interference to the radio telescope than the interference protection threshold, and where if anyone plans to establish any emitter, governmental or non-governmental, they must first apply to the national radio regulatory authority and inform the radio observatory of the relevant technical details and the latter may lodge objections and complaints within a certain period.

The radio quiet zone, whose size depends on the geographical environment and the transmission environment of the radio observatory, is generally applied by the observatory and legally determined and protected by the national radio regulatory authority and local government.

2.3   Radio Environment Monitoring for the Candidate Site of FAST

Scientists conducted sample monitoring in the 25–1500 MHz band in eight candidate depressions from 1994 to 1995, and the result showed that the depressions' interference level is less than one ten-thousandth of that of nearby towns. To protect this resource, the Chinese Academy of Sciences inked an agreement on establishing a local radio quiet zone in 1998. The observation results in the depressions obtained by the Office of the Radio Administration Committee of Guizhou Province in January and July of 2000 showed that the radio environment keeps stable.

In November 2003, the National Astronomical Observatories of the Chinese Academy of Sciences (NAOC) signed a cooperation agreement with the Office for 18 months. Guizhou Radio Monitoring Center, under the guidance and with the cooperation of NAOC, completed the measurement and international calibration of radio interference in key depressions in Guizhou in accordance with the requirements of the International Guidelines on Radio Frequency Interference. In September 2005, a five-week radio frequency interference monitoring was carried out in Dawodang Depression in Guizhou, with band coverage ranging from 70 MHz to 18 GHz. "The radio environment here is incredibly quiet and it is the ideal site for radio telescopes", said the head of the International RFI Monitoring Team.

Knowledge Link
RFI
  RFI refers to Radio Frequency Interference. For radio astronomy, the common sources of radio frequency interference include wireless, television relay, radar signal, etc.

2.4   FAST Radio Environment Protection

In order to protect the quiet radio environment around FAST site, Guizhou Province issued the Decree No. 143 of Guizhou Provincial People's Government in October 2013, *Measures for the Protection of the Radio Quiet Zone for the Five-hundred-meter Aperture Spherical Radio Telescope* (hereafter referred to as the *Measures*), establishing a radio quiet zone with a radius of 30 km with the FAST site as the center. The zone was divided into three subzones according to the different requirements for radio environment under the condition that the radio environment around the site is strictly protected and people in the surrounding towns and villages can live and work comfortably.

In September 2016, the *Regulations on the Protection of Radio Environment for the Five-hundred-meter Aperture Spherical Radio Telescope in Qiannan Buyei and Miao Autonomous Prefecture* (hereinafter referred to as the *Regulations*) were promulgated and put into effect, protecting the electromagnetic and ecological environment within a radius of 5 km around the site and providing a legal guarantee for FAST's radio quietness. A restricted airspace with a radius of 30 km was set up after two routes were relocated in August 2017, during which time the radio regulatory authority of

Guizhou Province provided staunch support in spectrum management, interference monitoring and investigation and other aspects.

While realizing FAST project's scientific objectives, it also matters to bring benefits to the society. In order to promote the project's role in facilitating the social and economic development in science and technology, education, tourism, information, manufacturing and other industries in the province, the provincial government organized and completed the compilation of the *Overall Plan for the Construction of Supporting Facilities in Guizhou Province for the Five-hundred-meter Aperture Spherical Radio Telescope*. As a result, the construction of the supporting facilities in areas such as transportation and tourism is underway. In the face of the coordination among many radio communication services and local economic development plans, how to strike a balance between scientific needs and the needs of local development is the core agenda of establishing the radio quiet zone.

Knowledge Link

Do not use cell phones near FAST site

FAST is extremely sensitive and vulnerable to external and internal radio interference. It can be interfered with by the use of cell phones, electrical devices, or by messages sent by an airplane in the sky to the ground nearby.

Therefore, the Radio Compatibility Working Group, in charge of the coordinating the radio interference protection measures, was established in December 2010 to protect FAST from radio interference, and to ensure its normal operation and scientific output. Guizhou province also enforced law requiring all base stations within a 5-km radius of FAST to be turned off, resulting in no cellphone signal around FAST. In addition, digital cameras around FAST are not allowed to be used, and cars cannot be ignited electronically.

3 Braving Hardships and Difficulties to Build the "Sky Eye"

The engineers for the FAST project completed the project after five-and-a-half-year hard work in September 2016 after braving hardships and difficulties such as the remote location, technical challenges and the large amount of non-standard equipment (special equipment).

Fig. 12 The FAST site in excavation

3.1 The Excavation of the Site and Slope Treatment

The excavation work of FAST site includes the removal of earth and rocks, slope treatment, and road and drainage system, so as to create basic conditions for the installation of various systems of the telescope, and provide a safe and stable external natural environment and infrastructure for the operation and maintenance of the telescope after its completion. It remains essential to continue to monitor the stability of the site, restore and conserve the soil and water, and prevent and control geological disasters after the completion of the excavation project.

Facing the complicated geological conditions and challenging construction conditions, the project team adopted a variety of measures to effectively manage the site during the process (Construction duration: March 5, 2011–December 31, 2012) (Figs. 12 and 13).

3.2 The Mounting of Structural Steelwork of the Ring Beam

The ring beam, the supporting system of the cable net of FAST, is composed of the basement of bearing platform, latticed columns, and the ring beam, By erecting 50 latticed columns with a height of 6.419–50.419 m, the ring beam with a diameter of 500.8 m, a height of 5.5 m and a width of 11 m was supported on 50 basements of bearing platform. Two aseismic bearings are installed on the top of each latticed

Fig. 13 In December 2012, the excavation and slope treatment of FAST site passed acceptance checks

column. The ring beam can slip in radial direction by slipping its bearings after expanding or contracting when subjected to changes in temperature. At the bottom chord of the ring beam which is with a circumference of about 1600 m, there are 150 node balls that pull the drag cable of the cable net. Each node ball is welded with an ear plate, and the center of each ear plate hole forms a circle with a diameter of 500 m. The construction of the ring beam was completed on December 31, 2013, marking the first milestone in the construction of the FAST project. On September 11, 2014, the engineering for manufacturing and mounting the ring beam passed acceptance checks (Construction duration: April 27, 2013–December 31, 2013) (Figs. 14, 15 and 16).

3.3 The Engineering for Manufacturing and Mounting the Cable Net

The cable net system is the main supporting structure of the active reflector of FAST, and the key point of the displacement working state of the reflector. The engineering for manufacturing and mounting the cable net is one of the main technical difficulties in the engineering of the 500-m spherical radio telescope. On August 6, 2015, the engineering passed the acceptance checks and the successful completion of it is also an important step of FAST project, bearing a landmark significance.

Fig. 14 The construction of the ring beam

Fig. 15 The structural steelwork of ring beam of FAST project was completed

Fig. 16 Aerial photo of the ring beam

The cable net of FAST is the cable net system with the largest span and the highest accuracy in the world, and also the first cable net system adopting the displacement working mode in the world. In 2015, The FAST engineering of the ring beam and cable net has obtained the special item of "Science and Technology Awards of China Steel Construction Society" (Construction duration: July 17, 2014-February 4, 2015) (Figs. 17 and 18).

3.4 The Engineering for Manufacturing and On-Site Mounting Feed Towers

The feed source supporting towers are the main bearing structure to support the feed support system of FAST, also the bearing and driving holder of the steel cables to provide an enough rigid supporting platform for the guiding pulleys on the top of towers and to assure that the driving steel cables can pull the feed cabin to move along the predicted trajectory. On November 30, 2014, the engineering for manufacturing and mounting feed source towers passed acceptance checks (Construction duration: March 15, 2014–November 15, 2014) (Figs. 19 and 20).

Fig. 17 The mounting of the cable net

Fig. 18 The engineering for the cable net system of FAST was completed in February 2015

Fig. 19 The feed source supporting towers

3.5   The Engineering for Manufacturing and Mounting the Cable-Driven System

As the largest cable-pulling parallel mechanism in the world, the cable-driven system is one of the three independent innovations of the FAST engineering. It is composed of the driving device, the guiding device, the cable device, the control system, the equipment base, and other ancillary facilities. On October 12, 2014, the on-site mounting of the cable-driven system officially kicked off. On February 10, 2015, the first supporting cable for the cable-driven system of FAST was successfully mounted. On November 29, 2016, the engineering for manufacturing and mounting the cable-driven system passed acceptance checks (Construction duration: October 12, 2014–May 18, 2015) (Fig. 21).

Fig. 20 The Panorama of feed source supporting towers

Fig. 21 The first supporting cable for the cable-driven system of the FAST was successfully mounted

3.6   The Engineering for Hydraulic Actuators of the Active Reflector of FAST

Under the control of the host control system, and through the piston movement of hydraulic actuators, the hydraulic actuators of the reflector realize accurate locating and congruous motion; through adjusting the lower-end positions of the down-tied cables, the node positions of the cable net are indirectly and synchronously adjusted, so as to realize the temporal paraboloid with a diameter of 300 m and the required high accuracy in real-time, and to meet the requirements of tracking, source exchanging, etc. during the astronomical observation. At the same time, the hydraulic actuator can also report its state information to the host control system according to the inquiry instructions from the latter.

On March 29, 2015, the first batch of 100 hydraulic actuators was manufactured and tested and left the factory. Since then, the production of hydraulic actuators is on the right track. The manufacturing, on-site mounting, and adjusting processes of all actuators were finished in July 2015 (Figs. 22 and 23).

3.7   The Manufacturing and Mounting of the Feed Cabin and Its Parking Platform

The feed source cabin is composed of a star frame, AB axes mechanical structure, Stewart platform, multi-beam receiver steering device, the protective cover of the cabin and other ancillary facilities. In October, 2014, the on-site mounting of the acting feed cabin was carried out in Dawodang, followed by the commissioning test with the cable-driven system. In March, 2015, the engineering passed the pre-acceptance checks. The acting feed cabin was mainly used in the early commissioning and experiment of FAST. At 11:00 a.m. on November 21, 2015, the feed source supporting system was successfully lifted for the first time, which marked that this system officially entered the phase of joint adjustment stage of on-load 6 cables. On February 17, 2016, the on-site mounting of the feed source cabin (focus cabin) began. On July 11, 2016, the feed source cabin (focus cabin) was officially lifted to 137 m.

The parking platform, located at the bottom of the center of the active reflector, is a platform for the mounting, parking, maintenance and testing of the feed cabin as well as the platform for the mounting and replacement of cables of cable-driven system. On November 30, 2015, the platform passed acceptance checks (Fig. 24).

Fig. 22 The mounting of
actuators

3.8   The Measurement Base and the Integrated Wiring System

The measurement base is the main building of the measurement and control system
for the FAST construction. 24 base piers that jut out over the reflector in the Dawodang
depression were built to provide a stable and reliable mounting platform for the high
accuracy measuring instruments, so as to perform the position measurement of the
reflector node points and the position/attitude measurement of the feed cabin and to
provide the measured data for the supporting and control of the reflector and feed
source (Construction duration: May 2013–October 16, 2014).

   The integrated wiring system (high and low voltage distribution of the internal
power grid, measurement and control network, and security and protection system)
is something like FAST's neural network, which is the channel for all instruction signals, data transmission and power transmission, and also the guarantee for
FAST's efficient operation. On June 25, 2016, the integrated wiring system engineering passed acceptance checks (Construction duration: April 2014–August 2015)
(Fig. 25).

Fig. 23 At the bottom of reflectors, yellow actuators are connected to the back frame by down-tied cables

Fig. 24 The acting feed source cabin was lifted for the first time

Fig. 25 The measurement base

3.9   The Development and Installation of Reflector Units

The reflector unit, like the lens of an optical telescope, is the panel of FAST that looks like a giant pot, and the pot has a spherical surface with an aperture of 500 m and a radius of 300 m. The reflector unit is connected to the main cable net node through the joints that are with certain degrees of freedom at the vertex, thus forming the reflector of FAST. Active displacement is the biggest characteristic of the reflector. A temporal paraboloid with a diameter of 300 m along the observing direction is formed through an active control to gather radio waves. During observation, the paraboloid moves on the spherical surface with a diameter of 500 m along with the diurnal motion of the observed celestial body, so as to realize tracking and observation. The project of the reflector is the last equipment project of FAST. After 11 months of arduous efforts, the construction personnel overcame the difficulties of large-scale and high-precision assembly construction, as well as the difficulties of large-span and high-position lift construction. The lift of the last reflector unit was completed in July 3, 2016, which marked the successful completion of the main engineering of FAST project (Fig. 26).

3.10   The Development of Receivers and Terminals Systems

The team of FAST project designed and developed 7 sets of receivers and terminals, including receivers, the time/frequency standard, the data transmission, the processing and storage, and the receiver monitoring and diagnostic system for the joint adjustment and scientific observation of the telescope.

Fig. 26 The construction personnel were lifting the last reflector unit

The role of feed source is to convert the spherical wave gathered by the reflector into the wave-guide and the guided traveling-wave in the coaxial line. Different feed source forms are adopted according to different frequency coverages of each band.

For the bands of 70–140 MHz and 140–280 MHz where the feed source has to work with high sky background noise, room temperature preamplifier and radio frequency circuit are adopted and the whole receiver works at ambient temperature. The 270–1620 MHz wideband feed source was developed by China in cooperation with California Institute of Technology in the United States, of which the feed source was independently processed and commissioned by China. The receiver was installed in September 2016. The 560 ~ 1120 MHz feed source adopted the corrugated horn and the polarizer adopted the form of broadband array plus waveguide. For the band of 1100–1900 MHz and 2000–3000 MHz, the feed source adopted corrugated horns and the polarizer adopted the form of quad-ridged waveguide. The receiver with frequency coverage of 1050–1450 MHz is a 19-beam multi-beam receiver, using 19 independent receiver units, each of which is composed of a receiver, a vacuum window, a thermal isolation waveguide, and a polarizer. In May 2018, the 19-beam receiver was installed, which meant that FAST's field of view in the L-band was expanded to 19 times of its original size, significantly improving FAST's sky survey efficiency (Figs. 27, 28 and 29).

Fig. 27 The low-frequency broadband feeds were installed in the lower platform of the feed cabin

3.11 The Construction of Observation Station

The construction of observation station is the basic guarantee for constructing, running, and maintaining FAST, which mainly includes two parts: (1) the basic construction of observation station, including the Complex Building, the Canteen and annex building, No. 1 Laboratory and No. 2 Laboratory; (2) public and ancillary facilities including road works, water supply works, power supply works, communication works, etc. The construction of observation station was completed on July 31, 2016 (Construction duration: October 19, 2015–July 31, 2016).

3.12 The R&D and Implementation of Electromagnetic Compatibility

FAST has a very high sensitivity and is very vulnerable to electromagnetic interference. Therefore, on the one hand, it is necessary to protect the quiet radio environment around the site; on the other hand, due to the complexity of FAST project, there are thousands of sets of electrical and electronic equipment, which puts forward high requirements for FAST's own electromagnetic compatibility, making special

Fig. 28 Picture of laboratory test of 19-beam receiver

electromagnetic compatibility design for FAST a must. In order to protect FAST from electromagnetic interference and ensure its normal operation and scientific output, the electromagnetic compatibility working group carried out a comprehensive and systemic electromagnetic compatibility design and successfully implement the design (Figs. 30 and 31).

3.13 Early Scientific Studies of FAST Project

In the five years before FAST went into operation, the Chinese radio astronomy community, by making full use of the existing equipment, achieved the world's

Fig. 29 L-band 19-beam
receiver

leading results in related fields and made scientific preparations for the official operation of FAST. Meanwhile, the scientific department of FAST was established and FAST project was supported by National Basic Research Program of China (973 Program) of the Ministry of Science and Technology of P.R.C. in a bid to train the radio astronomical talents and conduct the studies on the early scientific objectives, which was expected that some important scientific results will be obtained in the name of Chinese scientists after the construction of FAST.

After five and a half years of strenuous efforts, FAST project was completed on September 25, 2016 (Figs. 32, 33 and 34).

Fig. 30 The complex observation building

Fig. 31 The main control room of FAST

Fig. 32 The completed FAST Project

Fig. 33 The aerial photo of FAST

Fig. 34 A group photo of the engineering team after FAST is completed and put into operation

4 Technical Challenges and Breakthroughs in Design

4.1 Active Reflector

(1) Actuators

In order to realize the deformation of FAST reflector from a sphere to form a paraboloid, 2225 sets of hydraulic actuators are needed. Actuators can perform deformation movement and location of cable net according to control instructions.

Active reflectors of FAST project need 2225 actuators which need to run continuously during the period when the telescope observes celestial bodies. As the executor of the reflector to achieve real-time paraboloid, the actuators have extremely high complexity in their function and performance requirements and specialness in their working state, and they also require large quantity and load capacity as well as demanding continuous running time, which all pose great challenges to the current domestic and foreign industrial technology. Although faced with extremely limited budget and time, technicians made a series of breakthroughs in the aspects of highly reliable heat dissipation, precise position control, high shielding effectiveness, high maintainability, modular design, lightweight design, simplified design, design of fiber optic communication system, embedded control system, program downloading and so on. The technicians overcame the difficulties by breaking up the whole into pieces (namely by dissolving risks through using small and easily replaceable parts instead of large parts) and with generalized reliability (namely the design that takes the reliability of maintainability into consideration), significantly reducing the risk of the whole project, ensuring the completion of the project as scheduled, and laying a solid technical foundation for the follow-up commissioning, maintenance, optimization and upgrading of FAST. During the development and manufacture of hydraulic actuators for the active reflector of FAST project, the engineers applied for a total of 14 national patents (Fig. 35).

(2)   Cable Net

The cable net of FAST is a spherical cable net used for carrying triangular reflector units. Each node of the cable net is connected to a hydraulic actuator by a down-tied cable and the actuator is connected to a ground anchor on the ground. The cable net adopts the unique displacement working mode, that is, according to the orientation of the observed celestial bodies, a paraboloid with a diameter of 300 m is formed in different areas of the reflector with a diameter of 500 m after the down-tied cable is controlled by the actuator. In this way, the telescope is able to observe celestial bodies.

Fig. 35 The hydraulic actuator of FAST

The cable net of FAST is the cable net system with the largest span and the highest accuracy in the world, and also the first cable net system adopting the displacement working mode in the world. The engineering for manufacturing and mounting the cable net is one of the main technical difficulties in the engineering of the 500-m spherical radio telescope. The key technical problems include: the design of mounting the super-span cable net, the development of steel strand with ultra-high fatigue resistant property, and the manufacturing technology of cable structure with ultra-high precision, etc. The successful completion of the cable net engineering means that an essential breakthrough in the above-mentioned key technical difficulties has been made (Fig. 36).

With a diameter of 500 m, the cable net system is divided by geodesic grid and designed in a discontinuous way, that is, the main cable net is disconnected by nodes. Some key indexes of the cable net structure are much higher than the relevant domestic and foreign norms. For example, the control accuracy of the main cable segment shall be less than 1 mm, the position accuracy of the main cable node shall be 5 mm, and the fatigue strength of the cable component shall not be less than 500 MPa. There are 6670 main steel cables, 2225 main cable nodes and 2225 down-tied cables on the whole cable net, the total mass of which is over 1300 tons. There are 16 kinds of sizes of main cable sections, and the cross-sectional area is somewhere between 280 and 1319 square millimeters. Due to site constraints, all cable structures must be assembled high in the sky.

Fig. 36 The down-tied cables on the back of the reflectors are connected to a total of 2225 actuators

(3)   Reflector Unit

The reflector unit of FAST is designed according to the structure, form and size of the main cable net. The design side length of the ordinary triangular reflector unit is 10.4–12.4 m. Each reflector unit weighs 427.0–482.5 kg and has a thickness of 1.3 m. The reflector unit is placed on the cable net node disk through a connecting device at three endpoints.

FAST is used to receive remote, weak radio signals, and that is why it is located in Qiannan, Guizhou province where there is no radio interference. However, the wet and rainy climate there brings difficulties to the transportation and assembly of reflector units, requiring people to draw on the wisdom of all and innovate thinking for the design and development of reflector units.

First of all, in order to receive remote and weak radio signals, the surface shape accuracy of the reflector should meet certain requirements, that is, the root mean square error of each reflector unit should be equal to or less than 2.5 mm, and the surface shape should be a spherical surface with a radius of curvature of 315 m. For a triangular reflector unit with a side length of about 11 m, it cannot be a simple structure, but must have sufficient stiffness and strength and have a surface shape adjustment device if it is expected to overcome the influence of gravity and wind load and ensure that its surface shape meets requirements. Second, the antisepsis requirement of the reflector unit must be taken into consideration. Third, given that the side length of the reflector unit is about 11 m, the reflector unit cannot be manufactured into a whole and transported to the site, making on-site assembly, adjustment and testing a must. In addition, when FAST is tracking and observing celestial bodies, the cable net nodes move back and forth along the direction of the center of the sphere according to certain rules under the action of the actuator, which makes the reflector unit move accordingly. Therefore, there must be a corresponding connecting device at the endpoint of the reflector unit to be connected with the cable net node disk. Finally, the reflector unit should not be too heavy, or the forces on the cable net and the ring beam will increase and lead to the increase of construction cost.

Each reflector panel unit is composed of 100 riveted panel subunits which consist of aluminum perforated sheets, connecting disk, purlin, etc., and they are mounted on the 66 adaptive nodes above the back frame which has a permeability of greater than or equal to 50% (on the one hand, the thru hole can reduce the weight, on the other hand, its light transmittance characteristic is conducive to the growth of vegetation under the reflector unit).

In order to ensure the final surface shape accuracy of the reflector unit, for the riveted panel subunits of the basic type reflector unit, the deflection of the panel center shall not be greater than 1.5 mm when placed horizontally (Figs. 37, 38 and 39).

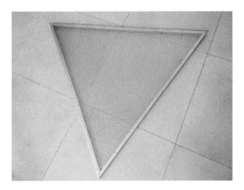

Fig. 37 Panel subunit

Fig. 38 The thru holes on aluminum perforated sheet

Knowledge Link

Will rain affect FAST?

In fact, precipitation, clouds and other meteorological factors have little influence on the transmission of radio waves with longer wavelengths, and only have some influence on the transmission of radio waves with shorter wavelengths. In addition, the working wavelength of FAST, especially the core working wavelength, is almost insusceptible to meteorological factors. Therefore, FAST can observe celestial bodies day and night and in sunny or rainy days.

Will FAST get flooded on stormy days? The answer is also no since the aluminum panels on reflectors are riddled with thru holes that allow rainwater to leak through, and the karst depressions where it is located also have a natural drainage system. In addition, a dedicated drainage channel was built below the reflector. Moreover, the vegetation below the reflector can grow as they want with sufficient sunlight, water and air, thus forming a reliable protection for rocks and soil at the bottom of FAST.

Fig. 39 Triangular mesh aluminum panel reflector unit and supporting back frame

4.2 Cable-Driven System

The Arecibo telescope adopts three huge steel cables to suspend a platform of nearly 1,000 tons. If FAST was going to be built this way, its platform would have a mass of 10,000 tons, which is not only expensive and difficult to construct, but also beyond the limit of present engineering technology.

Chinese scientists put forward a bold idea: using six light-weight steel cables to pull a huge astronomical receiving equipment platform—the feed cabin, to realize high-accuracy pointing and tracking of the telescope's receiving equipment by adopting optical, mechanical and electronic integration technology. The innovative technology has reduced the mass of the nearly 10,000-ton signal receiving platform to dozens of tons. From the perspective of theory of mechanism, this is called the flexible cable-pulling parallel mechanism, which falls to the category of cable-driven system engineering in FAST project, and this mechanism of FAST has the largest span among all completed projects in the whole world.

It is a challenging technical problem that the cable-driven system must realize the feed source cabin locating with a high precision and a large coverage. The cable-driven system engineering requires knowledge of more than a dozen of professional fields, such as astronomy, radio, machinery, electrical engineering, communications, measurement and control, posing challenges in technologies in terms of large span, high precision of flexible control, wide range of speed regulation, sophisticated process, and mounting difficulties. A number of technologies break through the existing standards and norms and there are no precedent examples to learn from. The Chinese scientists, amid the construction of FAST, developed for the first time in the world the moving fiber optic cable whose performance is far better than that required by Chinese military standards. At the same time, the scientists have innovatively developed curtain-like mechanism of cable into focus cabin which meets the transmission requirements of FAST for observing signals. As the device of cable-driven system includes high-power electronic devices and motion mechanisms, electromagnetic shielding measures that meet the highest requirements of military standards are adopted, of which many technical solutions are applied for the very first time.

The cable-driven system consists of 6 driving mechanisms, each of which includes a large motor with 257 kw power that drives the reducer with high reduction ratio and rotates the drum of the reel lock. In fact, the motor, reducer, drum, and brake together constitute a set of a common winch. It's just that FAST project requires higher precision and more power (Figs. 40 and 41).

There is a big pulley with a diameter of 1.8 m at the bottom and top of six towers which have a height of 100 m. The pulley at the top of the tower can rotate along with the motion direction of astronomical receiving equipment and its function is mainly to change the direction of steel cables. The flexible cable-pulling parallel robot is mainly driven by a driving mechanism installed in the machine room at the bottom of the tower to drag the steel rope to drive the feed cabin to move in the air. One end of the steel rope is fixed on the drum set of the driving mechanism, and the steel rope is led to the top of the tower by the guiding pulley on the bottom of the tower and the guiding pulley on the top of the tower, and then connected to the astronomical receiving equipment platform after being fixed by its anchor (Figs. 42 and 43).

Fig. 40 The joint adjustment of the cable-driven system in the factory

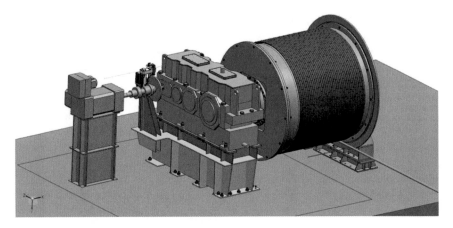

Fig. 41 Cable-driven system transmission mechanism

Fig. 42 The guiding pulleys on the top of towers

(1)   Moving Fiber Optic Cable

The most important feature of FAST's feed supporting system is the flexible connection between the feed source cabin and the ground. This flexible connection constantly changes the length and direction as the feed cabin moves in a wide range, which makes it difficult to establish a fixed connection mechanism of cabin into focus cabin between the feed cabin and the ground. The feed supporting adopting optical, mechanical and electronic integration technology is a brand-new design. The signal transmission between the telescope's ground control room and the feed cabin in the air is a key technology of the telescope, and the way it works is similar to that of the aerial cable used in conventional communications, but what makes it special is that the cable will expand or narrow continuously as the position of the feed source

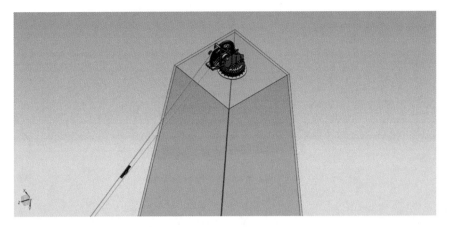

Fig. 43 The guiding pulleys on the top of towers and the curtain-like hanging-up mechanisms

is adjusted. Different from the relatively static working state of conventional cables, the cables used in FAST have to transmit analog signals in the process of repeated bending. At the same time, they must also have a very low loss to meet the signal monitoring channel of astronomical receivers. The protective coat of the cable not only should be able to withstand the harsh tests brought by natural environment such as the wind, sun, and rain, but also to overcome the impact of repeated bending and mechanical aging of each component of the cable in the process of movement.

In cooperation with universities and enterprises, FAST engineers and technicians, after four years' painstaking efforts, developed the moving optical cable whose fatigue life of 100,000 bending times is 1000 times of that required by the Chinese military standard, and the rate of change of signal attenuation less than 0.044 dB (1%) in motion state is 75% less than that required by the military standard. This achievement was made a special report in the 63rd IWCS International Cable Connectivity Symposium. As the first moving fiber optic cable in the world, it was officially promoted to the market in 2015.

(2)   The Transmission Technology for FAST to Observe Signals

FAST's feed source platform moves at a distance of 140–180 m above the depression, and a large-span flexible support structure is used to connect the feed cabin in the air with the control room on the ground. The project team innovatively developed the curtain-like mechanism of cable into focus cabin to establish power and signal transmission channels for the feed cabin suspended with 6 cables in the air. Thin steel ropes are used to pull the 86 pulleys suspended on each cable and when the thin ropes become longer, all pulleys will be rolled out and when the ropes become shorter, the pulleys will gather together at one end of the feed source platform. Those pulleys should not only be strong and resistant to corrosion caused by the acid rain in Guizhou, but also light in weight or it would undermine the control accuracy of the position/attitude of the feed source platform (Fig. 44).

Fig. 44 The curtain-like mechanism of cable into focus cabin

(3)   Electromagnetic Compatibility (EMC)

The radio telescope is demanding on the radio environment. When the telescope equipment runs, the equipment itself will produce various radio signals, which will

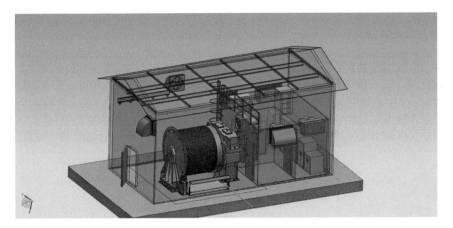

Fig. 45 Cable-driving shielding room

cause serious interference and even damage to the astronomical receiving equipment of the telescope.

The equipment includes high-power electronic devices and motion mechanisms, which will release radio waves during operation, making shielding difficult. Though a shielding room is built in the machine room and the equipment is installed in the shielding room to prevent the leakage of radio waves generated by the equipment, the leakage of radio waves is inevitable as the thin steel rope must go across the machine room through an opening and the motor will pass through the wall panel of the machine room during transmission. Finally, the project team adopts a labyrinth structure which is added to the drive shaft of the motor, effectively attenuating the radio wave and preventing its leakage (Fig. 45).

(4) Great Difficulties in Installation

The individual equipment of the cable-driven system has a maximum mass of 13 tons and needs to be installed in a machine room which is located on a hill 30 m higher than the road. The huge slope makes large-sized lifting equipment inaccessible. To tackle the tricky issue, the project team built a drag track and used a winch to haul the lifting equipment up to the hill to the machine room.

The main technical difficulties in mounting the supporting cable lie in: the steel rope needs to be mounted over a large span with a height difference of up to 277 m; the individual steel rope has a length of more than 600 m with a diameter of 46 mm and the mass is up to 6 tons; the horizontal span is 300 m. Under each of the 6 steel ropes, a cable with a diameter of 26 mm is suspended. Among the 6 steel ropes, 3 have a 48-core fiber optic cable with a diameter of 12 mm suspended below them respectively. That is why the unique curtain-like mechanism of cable into focus cabin is developed to suspend the cable and fiber optic cable. The mechanism of cable into focus cabin has many kinds of components and complex working conditions, which requires extremely high reliability. According to working conditions and force conditions, each steel rope should be equipped with 4 types of pulleys with a total of 86 sets.

Limited by factors such as the mass of steel ropes, the large span, sophisticated mounting of pulleys, cables and fiber optic cables, confined construction space and other complex working conditions, the contractor in charge of the construction, after repeated studies and experiments, finally decided to use both thick and thin ropes. With the thin rope hauling the thick rope and the thick rope hauling the steel rope, the double reverse twisted steel rope with a diameter of 46 mm is gradually drawn from the bottom of Dawodang through the guiding device at the top of the tower to be sealed on the drum in the machine room. During this process, 86 sets of pulleys, cables and fiber optic cables are installed in place with the steel rope. After the installation, the cable-driven system tested the communication of the 48-core fiber optic cable that was developed after 4 years of hard work and the smooth signal transmission was achieved successfully.

4.3   The Feed Cabin

As the core component of FAST project, the feed cabin is a multivariable, nonlinear and complex coupling multi-body dynamic system integrating structure, mechanism, measurement, control and other related technologies. It has a diameter of 13 m and a mass of about 30 tons. Its main function is to realize the accurate positioning of the feed after overcoming the wind disturbance and other disturbances that are under the influence of suspension cables. The two main bodies in the cabin are the AB axes mechanical structure and the Stewart platform (Fig. 46).

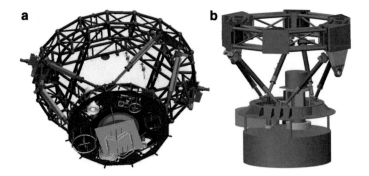

Fig. 46 Stewart platform

Knowledge Link

AB Axes

AB axes are bidirectional rotators that can rotate in two directions.

Stewart Platform

Stewart Platform is a typical parallel mechanism, and it is also a standard term. The platform is able to realize six degrees of freedom with the piston movement of its six retractable legs.

The feed supporting system includes both coarse positioning and precise positioning. Coarse positioning is achieved by connecting the three supports of anchor heads of the feed cabin by six steel cables, each of which has a length of hundreds of meters, and the cables drive the feed cabin which has a diameter of 13 m to move on the focal plane that has a span of about 206 m, and, together with the AB axes mechanism, realize the astronomical trajectory planning of the feed source. The accuracy of coarse positioning is up to 48 mm. In order to meet the requirement of astronomical observation and that of FAST's construction task, one Stewart platform is used in the feed cabin as a fine tuning positioning mechanism to reduce and suppress the influence of wind disturbance on the feed source positioning, which further improves the positioning accuracy of the feed source to within 10 mm of the root mean square value, so as to meet the accuracy requirement of astronomical observation.

The design of the feed cabin has gone through stages such as the plan of co-control by both the cable and the Stewart platform, and the cable-and-car plan, etc. It was not until early 2008 that the existing plan of both AB axes and Stewart platform as the main structure was basically determined. In 2009, with the financial support from the National Astronomical Observatories of the CAS, Tsinghua University built a feed supporting system model on a scale of 1:15 in Miyun District, Beijing. This model realized the main functions of the feed supporting system that we see now, and the terminal control accuracy reached 1 mm. In 2010, the University continued to design the feed cabin in order to refine its structure and finished the design in September, 2011, finally determining the form of the main structure of the feed cabin.

After half a year of optimizing the design, the project team embarked on the detail design phase for the feed cabin. After investigation and selection, the 54th Research Institute of China Electronics Technology Group Corporation (CETC54) was finally identified as the general contractor for the design, manufacturing, mounting and commissioning of the feed cabin, since then the R&D and construction work for the feed cabin officially kicked off.

(1)   The Components of the Feed Cabin

The components of the feed cabin are shown in Fig. 47. As the main bearing structure, the star frame is connected with the cable-driven system by 3 groups of anchor heads. It is installed with equipment for dynamic monitoring, lightning protection, power distribution, receiver, control cabinet, maintenance, firefighting, lighting and others. The AB axes mechanism rotates around the two orthogonal axes and realizes the

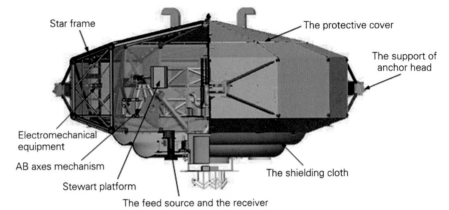

Fig. 47 Structure of the feed cabin

preliminary adjustment of the feed source attitude through the control system. The Stewart platform carries the feed receiver and adjusts it precisely through the control system. The multi-beam steering device is installed on the Stewart platform to realize the axial rotation of the multi-beam receiver. The protective cover is fixed to the star frame for the isolation and electromagnetic shielding of the equipment inside the cabin from the external environment. The wiring system includes the winding mechanism and cables inside the cabin. The power distribution system provides power for each system. The monitoring system includes temperature, humidity and smoke sensors, cameras and so on, which transmit the status of the cabin to the ground system. When there is equipment emitting electromagnetic wave with excessive levels of interference, the electromagnetic compatibility unit will play its role by shielding, filtering and ground connection.

According to the picture, it seems that the feed cabin with a diameter of 13 m doesn't stand out in the FAST that boasts a diameter of 500 m, and it often takes extra efforts to find it from the photo. However, the feed cabin is a complex component with many interfaces, such as cable-driven system, parking platform, receiver, measurement and control equipment, etc., which also brings such restrictions and requirements to its design as the mass and size restrictions, high standards of electromagnetic compatibility and complex control schemes.

(2)   Restrictions of Mass and Size

The feed cabin is towed to an altitude of 140 m by cable-driven system which, together with AB axes, jointly realizes the trajectory planning of the cabin. According to simulation analysis results and after taking factors such as cost, size, performance, and safety into consideration, the mass of the feed cabin should not exceed 30 tons.

According to the initial design requirements, the Stewart platform in the feed cabin would need nine sets of feed source, cover the band of 70 MHz–3 GHz and have a diameter of 13 m and a total mass of 30.7 tons.

Knowledge Link

The design thinking of the feed cabin

Installation requirements of feed source—the lower part of Stewart platform—optimal design of Stewart platform—AB axes mechanism—star frame. Based on this overall design thinking, there are some factors that should be considered as well, such as the measurement interface between the lower platform of Stewart platform and the star frame, the connection interface between the star frame and the cable-driven system, the interface between the electromechanical equipment in the star frame and where it is installed, and the interface of wiring system in the cabin.

The mass of the feed cabin has continuously increased until it hits 34 tons as the design has been optimized and detailed. The increased mass increases the force exerted on a single rope driven by the cable and reduces the safety factor of the rope, which is not conducive to the long-term safe use of the cable-driven system. At this point, as the design of the feed source and the receiver is further improved, the frequency band of the original 9 sets of feed source is redistributed, making 7 sets of feed source become a better choice. The change leads to the adjustment of the feed cabin as the characteristics and observation requirements of the 7 sets of feed cabin change. The one lower platform of Stewart platform is changed into two smaller lower platforms and AB axes and star frame also undergo some changes, which significantly reduces the mass without compromising the original structural rigidity.

Finally, the mass is reduced to 29.8 tons under the condition that the interface between the feed cabin and the cable-driven system is not changed.

It is worth mentioning that while designing the structure, the project team planned to apply bolted connection in the main structure of the feed cabin given that the on-site installation is easier. However, constrained by the mass, the designers had to abandon this installation method, turning to welding instead. Nevertheless, due to the high requirement of electromagnetic shielding and special requirement for the welding of the protective cover of the cabin, a lot of on-site welding work was inevitable, which also put forward higher requirements for tooling and welding and increased the construction difficulty (Fig. 48).

(3)   The Feed Cabin's High Requirement in Electromagnetic Compatibility

The characteristics of the single aperture radio telescope require very sensitive signal receiving system, which means that the electromagnetic interference produced by the equipment around the telescope must be less than the receiving capacity of the equipment itself, especially the electrical equipment nearby feed source. The electromagnetic shielding requirements of FAST project are far higher than those of China's military standard GJB151A, so more efforts and cost must be invested.

In accordance with the Recommendation: *Protection Criteria Used for Radio Astronomical Measurements of ITU* (ITU-R RA.769), it is theoretically estimated

Fig. 48 The shape of the
main body of the feed cabin

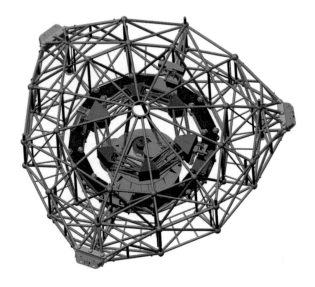

that, within the frequency range of 70 MHz–3 GHz, more than 160 dB shielding
effectiveness should be obtained after taking the equipment in the feed cabin and the
structural features of the cabin into consideration. To realize the index, the whole
structure of the feed cabin is shielded by steel plates, and shielding measures are also
taken for places where electromagnetic leakage is likely to occur, so that the whole
cabin becomes a seamless shielded conductor. As far as the equipment in the cabin
are concerned, shielding compartment is applied. Two-stage shielding measures are
adopted for some equipment with high electromagnetic radiation. The standard of
electromagnetic shielding effectiveness falls into two categories: 80 dB for outer
shielding of the feed cabin, and 80 dB for inner shielding of the feed cabin.

According to the design, the whole feed cabin is shielded by steel plates: the outer
surface of the star frame is welded with the stainless-steel plate with a thickness of
0.8 mm, and the plate is seamlessly connected with the star frame to avoid electromagnetic leakage. The overall shielding of the feed cabin is made possible by
using the protective shell on the surface of the feed cabin. The stainless-steel plate
(0.8 mm thick) is adopted to seal the cabin into a shielded enclosure, only leaving a
door on the outside for maintenance workers. The whole shielded enclosure of the
feed cabin is divided into two relatively independent compartments. Shielding steel
tubes are welded between adjacent shielding compartments for cable connection. A
door in each compartment is needed for equipment installation. Waveguide ventilation windows and forced exhaust equipment were added accordingly for ventilation
purpose. The lower platform of Stewart platform is in constant motion in the feed
cabin. To this end, a double-layer shielding and waterproof cloth is adopted on somewhere from the star frame to the lower platform of Stewart platform, which ensures
shielding effectiveness and environmental adaptability (Fig. 49).

According to the characteristics of electromagnetic interference, the equipment
in the cabin is reclassified into the following categories: the power supply and signal

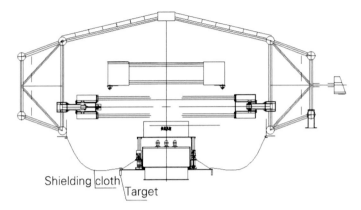

Fig. 49 Part design of the metal body of the feed cabin

circuit; the servo drive; the mechanical drive system (electric motor, encoder, etc.); other equipment (GPS, camera, etc.).

Outlet and inlet wires of power supply should be used in a power filter whose shielding effectiveness should be more than 160 dB for the outside of cabin and 80 dB inside the cabin. For the signal cable in the cabin that runs through the wall and thus cannot use the power filter such as motor wires, camera signal wires and control circuits, shielding wire tubes and ferrite beads will be used to wrap the wires to meet the requirement (Fig. 50).

The servo drive of the electric motor is installed in the shielding compartment of the star frame, so shielding measures must be taken for the signal cable and the power cord of the motor control system to run through the cabin. Scientists adopt different measures for different devices. For example, for AB axes, the original shielding plan of transmitting electromagnetic interference directly to space is changed into making a shielding can for the motor. Specifically, the motor would be placed inside the shielding can which, by bolt connection, is connected with the reserved shielding interface at the flange where the motor is installed. The conductive gasket is used between the shielding can and the flange to meet the shielding requirement (Fig. 51).

As for the camera, the radiation resistant camera system made by professional manufacturers is applied, and its radiation shielding performance in the range of

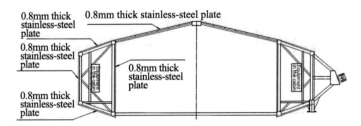

Fig. 50 The shielding design for the motion platform of the feed cabin

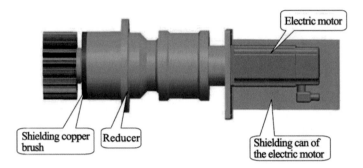

Fig. 51 The design for the shielding of the electric motor

14 kHz–18 GHz hits 80 dB. In terms of GPS, a professional signal isolator is used to prevent the signal from leaking out of the cabin.

(4) The complex control plan of the feed cabin

The positioning of the feed support system relies on the cooperation between the cable-driven mechanism, the AB axes, and the Stewart platform. The feed cabin will vibrate mainly under the influence of wind force and other external factors, so the Stewart platform needs to be controlled to inhibit the pose change of the lower platform under external influences, so as to meet the requirement of the feed's positioning accuracy of 10 mm. In general, such requirement is not daunting for parallel mechanisms like the Stewart platform, but as the precision control of the feed cabin becomes complicated by the flexible support of the cable-driven mechanism, the control coupling of rigid and flexible mechanisms should be taken into full consideration, and the feed cabin vibrations caused by the counterforce of the driving force of the Stewart platform acting on the feed cabin should be analyzed.

With regard to the control of the feed cabin, on the one hand, scientists proposed effective control plan and extracted control parameters based on their understanding of the features of cabin cable system derived from simulation analyses. On the other hand, they measured the pose of the lower platform through Electronic Total Station (ETS), for the sake of closed-loop control and improvement of the control precision (Fig. 52).

The Stewart mechanism has a load of nearly 3 tons, the total mass of the feed cabin is about 30 tons, and the damping ratio of the cabin cable system stands at 0.2%. Subsequently, scientists conducted modal analysis and frequency response features analysis of the cabin cables.

The Stewart platform adopts joint space control strategy. Operating in the joint coordinate system, it is the controller of each of the driving legs of the Stewart platform. By enabling each and every driving leg to track precisely the expected length of driving legs calculated through inverse kinematics, it ensures the overall performance of the Stewart platform.

In actual control, the pose of the upper platform of the Stewart platform is realized through the control strategy algorithm.

Fig. 52 Structure model of the feed cabin

One disadvantage of this control strategy is the issue of structural deformation. In the control of the Stewart platform in the past, it was generally believed that its stiffness is similar to that of a rigid body. The structural deformation mentioned in the past usually fell in the tolerance range of the structural components, and such structural errors can mostly be compensated through calibration, leaving relatively small remaining errors. However, due to the low stiffness of the feed cabin, coupling relationships will exist between the structural parameters in actual movement control, so that the influences of the structural deformation on precision, which cannot be reduced greatly through calibration, will be reflected more in the actual control process. Therefore, some part of the control errors will be compensated through the current terminal measurement method and the control strategy, while for others, the control parameters need to be corrected through long-term accumulation of operation data (Fig. 53).

4.4 Measurement and Control

(1)   Benchmark Network and Basic Related Measurements

A high-precision and permanent GPS benchmark control network has been established surrounding the Dawodang depression, serving as the primary geodetic control for surveying and mapping the large scale topographic map of the FAST site and the engineering design graph. This network can provide point coordinate benchmark, length benchmark and time benchmark for the work of construction positioning surveying, construction lofting and operation of the feed support towers and the main reflector's ground anchor. At the same time, it can also constitute the ground reference

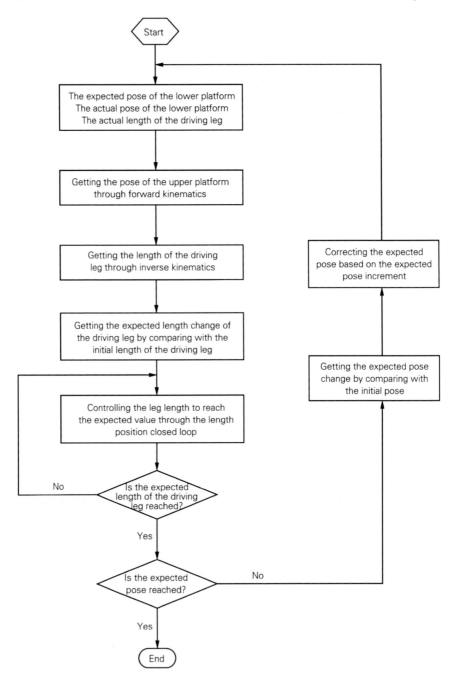

Fig. 53 Flowchart of the control strategy algorithm of the Stewart platform

Fig. 54 Layout plan of the benchmark network. *The horizontal axis represents the east–west direction of spatial position, and the vertical axis represents the north–south direction

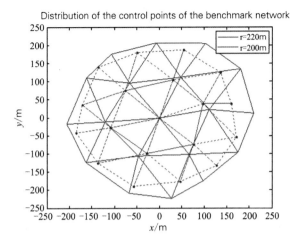

Distribution of the control points of the benchmark network

station for the real-time feed tracking, positioning and measurement system as well as the active deformation control of the reflector. The basic measurement subsystem provides the coordinates of the fixed point of the main cable and the down-tied cables and of the position of the feed support towers and the cable outlets (Fig. 54).

The optimized result of the 220-m-radius benchmark network is finally selected, considering that it has wider coverage and that there are no significant differences in the height of corresponding base piers (Fig. 55).

(2) Feed support measurement

On the 600-m-diamter circumference surrounding the reflector, six 100-m-high towers support six steel cables, which drag the FAST feed cabin to move within an aperture of 206-m diameter at the altitude of 150 m, and also complete high-precision astronomical tracking movement simultaneous with the reflector in accordance with the astronomical planning and measurement information feedback. The AB axes mechanism and the Stewart platform parallel mechanism inside the feed cabin are used for the precise positioning and tuning of the astronomical information receiving system, which is installed on the Stewart platform for the collection and transmission of astronomical information.

The feed support measurement system consists of the pose measurement system of one-time cable-driven mechanism and the pose measurement system of the fine tuning platform. It utilizes measurement instruments with high precision to provide precise pose information of the one-time cable-driven mechanism and of the fine tuning platform for the feed support control system, so as to complete the task of measuring the pose of the feed, and realize the precise positioning of the feed.

Once started, the measurement system begins its initializing operation, finishes the initial positioning of the feed cabin through the GPS measurement system, and then goes further to complete the orientation and target locking of the Electronic Total Station (ETS). After being initialized, the system goes into standby status,

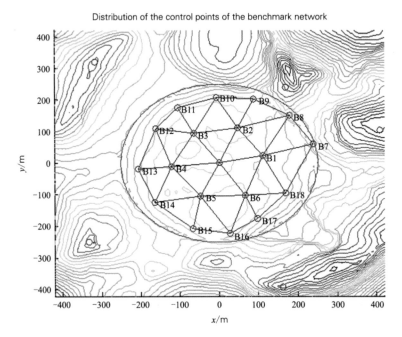

Fig. 55 Serial numbers of base piers and their corresponding control points. *The horizontal axis represents the east–west direction of spatial position, and the vertical axis represents the north–south direction

waiting for the instruction of measurement. Different working modes, including the observation mode and the debugging mode, are chosen according to different measurement instructions. After completing the assigned work, the system returns to the standby status and waits for new measurement instructions.

①   Pose measurement of the one-time cable-driven mechanism

The pose measurement of the one-time cable-driven mechanism is composed of the GPS and the ETS measurement instruments. Among them, the design indicators of the GPS measurement system include the precision of 2 cm and the sampling rate of 1 Hz. The measurement system mainly comprises reference stations and moving stations, and utilizes double reference stations. The GPS receivers located on the measurement reference points at the periphery of the depression are regarded as the reference stations which provide differential reference for the measurement system. At the same time, they serve as the backup for each other, so as to enhance the reliability of the system. The moving stations are the six GPS receivers installed on the top of the feed cabin.

The measurement equipment comprises the three ETS instruments located on the reference points within the reflector and the three prisms located on the fringe of the feed cabin. The coordinates of the measurement points are obtained through real-time dynamic measurement with a maximum error of 17 mm.

Knowledge Link
Electronic Total Station (ETS)

Electronic Total Station is a high-tech photoelectric measuring instrument with optical, mechanical and electrical integration by combining electronic theodolite, photoelectric distance measuring instrument, and microprocessor. Such surveying instrument system integrates the measurement of horizontal angles, vertical angles, distances (sloping and horizontal), and height differences. In comparison to optical theodolite, electronic theodolite replaces optical circle with photoelectric scanning circle, and replaces manual reading of optical micrometer with automatic record and display of reading, which not only simplifies the operation of angle measurement but also avoids reading errors. It is called ETS because with one-time installation it can complete all of the measurement work on the measurement station. ETS embraces incredible precision in terms of angle and distance measurement. Operable both manually and automatically, and both through long-distance remote control and through control by instrument-end applications, it has already been applied to precise project surveying, deformation monitoring and other fields.

② Measurement of the fine tuning platform

The fine tuning platform measurement plan adopts the ETS measurement system. Several ETS instruments are installed on several base piers within the reflector. Six measuring targets are installed evenly on the lower platform of the fine tuning mechanism in the feed cabin. The ETS instruments are targeted at the corresponding measuring targets to measure and then obtain real-time dynamic measurement data of distances and angles.

Influenced by the refraction of the laser emitted by the ETS instruments in different atmospheric environments, there might be errors in the measurement data. Distance measurement can be corrected through models, after which higher precision can be obtained; while angle measurement lacks reliable and precise correction models, so in the fine tuning platform measurement, after correcting atmospheric parameters of the distance measurement, the position and attitude of the lower platform will be calculated through the distance intersection method.

③ Electromagnetic interference (EMI) shielding of the measurement equipment

What FAST observes are extremely remote and weak radio signals in the universe, which are extremely sensitive to the influences of radio frequency interference (RFI). So it is quite significant to control the radio frequency radiation of the various kinds of electronic products used in the telescope project.

The ETS instruments are of crucial importance in the feed support measurement and the reflector node measurement. Besides, they are exposed above the reflector. Therefore, under the premise of not influencing the measurement precision of the

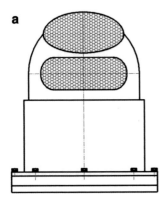

Fig. 56 ETS shielding design

ETS instruments, shielding materials are used for all-round EMI shielding of the ETS instruments, and the index of shielding effectiveness is 100 dB (Fig. 56).

(3) Reflector measurement

When the FAST reflector deforms into an instantaneous paraboloid with an aperture of about 300 m and the driving motor drags the nodes of the cable net, due to the huge structure of the cable net which leads to complicated force conditions, the nodes may traverse tangentially rather than move strictly in the radial direction. Besides, it is hard to estimate accurately the flexible deformation of the down-tied cables which bear enormous pulling force, which makes it not accurate to identify the positions of the nodes merely through the adjustment of the driving motor. Therefore, the real-time and accurate measurement of the nodes of the cable net is required to ensure the operating precision of the telescope.

Unlike the reflectors of other telescopes which only require regular surface shape detection, the FAST reflector goes through real-time deformation during observation, which means real-time precise measurement and control is the key for FAST to achieving excellent observation performance. However, with no reflector measurement cases from other telescopes serving as the reference for this technology, requirements of high precision and high efficiency, real-time interaction with the control system, field environment, atmospheric interference, long-distance measurement, and the like, all constitute the technical difficulties and bottlenecks in reflector node measurement.

After a large amount of early stage investigation, the reflector node measurement system adopts the method of using several laser ETS instruments to measure point by point. This method shows great reliability and relatively mature technology, and the use of passive targets leads to less electromagnetic interference.

The FAST reflector node measurement system is able to carry out static and dynamic measurement of the nodes of the whole reflector net. It has two working modes.

Calibration mode: to conduct precise calibration of the positions of the 2225 nodes within the whole reflector when they are static. The measurement of all nodes of the whole net can be completed within 37 min, and the root mean square error of precision is less than 1.5 mm.

Observation mode: to measure the positions of the nodes within the instantaneously illuminated aperture of the reflector during the observation of the telescope. The measurement can cover approximately 700 nodes inside the effective aperture within 9 min, and the root mean square error of precision is less than 2 mm.

The FAST reflector measurement system mainly consists of base piers, the ETS instruments, and targets. The targets are installed on the reflector nodes, i.e. the junctions of the triangular panels. When the reflector is operating, the ETS instruments installed on the base piers are used to measure the positions of the targets inside the illuminated area, so as to control the reflector surface shape in a real-time manner.

① Measurement base piers

FAST has 24 stable measurement base piers, among which the five base piers near the reflector center belong to inner circle base piers (JD1–JD5), and there are six middle circle base piers (JD6–JD11) and 12 outer circle base piers (JD12–JD23). Moreover, another base pier (JD24) has been built on the Bright Summit of the mountain near the telescope (Fig. 57).

Three forced centering plates are installed on the top of each base pier for the simultaneous use by three instruments.

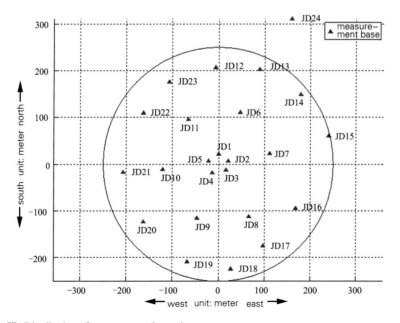

Fig. 57 Distribution of measurement base piers

Fig. 58 Leica high-precision laser ETS instrument TS60

② Laser ETS

The FAST reflector measurement system uses Leica high-precision laser ETS instrument TS60. As the world's highest-precision ETS instrument at present, it can work both in the day and at night (without illumination) owing to its Auto Targets Recognition function (Fig. 58).

③ Targets

In the measurement system, the targets used in coordination with the ETS are reflecting prisms, which are connected with the reflector nodes by screw threads. The role of the prism is to reflect the electromagnetic waves (laser) emitted by the ETS back to the ETS. The receiving device of the ETS receives the waves, and its timer can record the time difference between the emission of the electromagnetic waves and the reception, so that the distance between the ETS and the prism can be obtained. At the same time, according to the turning angles of the ETS camera lens, the horizontal angle and pitch angle between the ETS and the prism can be calculated, and thus the position information of the prism can be gained.

The prism used by the FAST reflector measurement system is shown in Fig. 59. According to the program design of the measurement system, when being installed, the targets should be pointed toward the measurement base piers in the center of

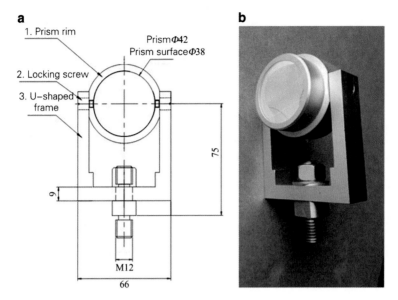

Fig. 59 Reflecting prism (target)

the reflector, and the pitch angle should be adjusted to the assigned degree and then locked.

FAST has 2,225 such prisms, which can work in coordination with the ETS instruments day and night to complete the measurement work. Figure 60 is a photo of a prism installed on the reflector node plate of the telescope; Fig. 61 shows the scenery

Fig. 60 Installed prism

Fig. 61 Scenery of the prisms photographed at night

of the reflection of the prisms on the reflector at night, when they are illuminated with light from the bottom of the center of the telescope.

4.5   Receiver and Terminal Function

We refer to feed, low-noise amplifier and receiver circuit, all the way to data processing terminal altogether as the receiver and terminal system. The following section is going to introduce the key technological research related to the receiver and terminal system.

(1)   Feed simulation design

Celestial bodies, the sources of radio radiation, are usually distant from the Earth. The electromagnetic waves emitted from the point of the position of a celestial body belong to spherical waves, whose equiphase surface is a spherical surface with the emission point as the spherical center. For telescopes on the ground, such spherical waves have subtle differences compared to plane waves, so they can be regarded as the latter. When the incident plane waves along the principal axis of the paraboloid turn to spherical waves after the reflection of the paraboloid, the spherical center of the spherical waves is the focus of the paraboloid.

The plane waves intercepted by the radio telescope reflector turn to spherical waves after the reflection of the paraboloid. The spherical waves focus the energy of the plane waves intercepted by the paraboloid on a very small area, which makes it possible to use a small detector to detect this part of energy. Unlike photographic films or films or photosensitive elements used in conventional optical astronomy, the feed is often used to receive the focused spherical waves in the radio band. The feed is a specially-shaped electronic device made of metal and other conductive materials. Figure 63 shows the FAST feeds developed for the L band and S band. There is no active component driven by power supply inside the feed, so it responds completely passively to the incident electromagnetic waves. The same goes for the reflection of the incident electromagnetic waves off the reflector and for the refraction of light through spectacle lens; they all adopt passive mechanism. To be specific, affected by the incident electromagnetic waves, the electrons inside the metal move at an accelerated speed and emit electromagnetic waves. In such case, the emission of electromagnetic waves by the metal materials of which the feed is composed is called secondary emission. The superposition of the secondarily emitted electromagnetic waves and the incident electromagnetic waves in space form the distribution of electric field in space. Well-designed feed will make electromagnetic waves appear to be smooth transition from spherical waves in free space to guided waves inside the feed space. The above-mentioned description shows the process of the feed receiving electromagnetic waves. If played back, the process becomes the transition from guided waves inside the feed to spherical waves in free space, realizing the emission of energy from the feed to the free space outside, which is one of the sources of the term "feed" (Figs. 62 and 63).

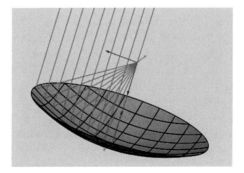

Fig. 62 Focus of the FAST reflector

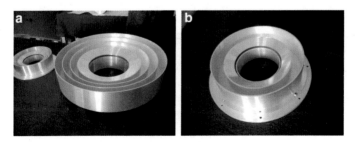

Fig. 63 The FAST feeds for L band (left) and S band (right)

(2)   The telescope far field pattern

The main reflector of a telescope is usually a paraboloid. The incident plane electro-magnetic waves along its main axis are reflected off the paraboloid and form focused spherical waves. When the feed is located on the focus of the paraboloid, its recep-tion of the focused spherical waves means that the radio telescope completes the observation of remote radio sources. The parabolic reflector and the feed located on its focus constitute a system which also responds to different degrees to the off-axis incident electromagnetic waves. The responses to the incident plane waves coming from different directions make up the far field pattern of the radio telescope, which shows its responses to incident plane waves from different directions. The biggest point, or the biggest response, usually appearing in the direction of the main axis of the parabolic reflector, is called the axial gain of the telescope. For a given obser-vation frequency and feed, the bigger the aperture of the reflector is, the larger the receiving area is, and the bigger the axial gain is. The axial gain reflects to some extent the size of the receiving area of a telescope.

(3)   Optimization of sensitivity

Sensitivity is an important performance indicator of a single-dish radio telescope. It determines the strength of the weakest radio sources that can be observed by the radio telescope. As mentioned above, radio sources are often rather far from us, and the electromagnetic waves received by the radio telescope can be regarded as plane waves. After being reflected by the parabolic reflector, the plane waves converge on the focus of the paraboloid, and are then received by the feed. The feed converts the electromagnetic waves transmitted in free space into guided waves transmitted in waveguides, the signals of which are processed through low-noise amplifiers, wave filters, and the follow-up radio frequency amplifiers, and are finally transmitted to the data processing terminal for the required spectrum analysis, dispersion correction, baseband record and other ways of processing. While receiving the radiation of radio sources, the radio telescope also receives sky background radiation, atmosphere radi-ation, ground leakage and other radio noises, and the aforementioned reflection of the reflector, the feed reception, and the low-noise amplifier will add extra noise signals to the received radio source radiation. Noise signals are different from the signals emitted by the cooperative targets, and we cannot predict the correlation between their specific voltage value and time. But noise signals present certain statistical laws, that is, their average power is generally relatively stable within a certain time. The root mean square of the average power of the noise signals is inversely proportional to the square root of the product of the bandwidth of the observed signal and the integration time. In general, the aforementioned noises caused by sky background, atmosphere radiation, ground leakage, and receiver noise are called system noise, which is represented by system noise temperature.

For a certain radio source, the effective receiving area of a radio telescope deter-mines the energy of the electromagnetic waves that can be received. As to FAST, its core observation band is L band, which has an effective receiving area of 36,400 m^2, the system temperature of around 25 K, and the sensitivity of about 1455 m^2/K.

The FAST main reflector measures 500 m in diameter with a spherical surface as the neutral surface, and its active deformation reflector measures 300 m in diameter. During observation, the large range of spherical metal plates outside the paraboloid can effectively shield the heat radiation from the ground, and thus lower the entire system temperature of the telescope. In the meanwhile, the shielding of ground heat noise by the spherical metal reflector outside the paraboloid also provides new possibilities for the optimization of feed illumination. Therefore, the lighting level of the feeds on the edge of the paraboloid can be widened, so as to maximize the ratio of the effective area of the telescope to the system noise temperature, and to obtain the optimal sensitivity. When the zenith angle is bigger than 26.4 degrees, scientists proposed the "back-illumination" approach of making the feed rotate inward around its phase center, in order to optimize the sensitivity of the telescope. After the zenith angle is bigger than 26.4°, some part of the 300-m active deformation paraboloid already stretches outside the ring beam of the 500-m-diameter telescope, so the area of the paraboloid available will be narrowed. Due to the fact that the feed is designed for the 300-m paraboloid, the feed located in the position of the truncated part of the paraboloid will receive heat radiation from the ground directly, leading to the increase of system noise temperature. In the early stage, the FAST project team considered the idea of building a plane metal noise-proof wall surrounding the periphery of the 500-m ring beam. However, in order to effectively shield the heat radiation from the ground even when the zenith angle is up to 40°, a metal noise-proof wall of approximately 50 m high is needed. Such a metal noise-proof wall per se is a large-scale infrastructure project which is very costly. The current design adopts the "back-illumination" model to optimize the analysis of sensitivity. With the "back-illumination" model, the symmetric axis of the feed far field pattern deviates from the parabolic center, but the aperture plane of the reflector corresponding to the illuminated area near the feed far field pattern is widened, therefore, the axial gain of the telescope will not undergo remarkable changes. However, after the "back-illumination" model is adopted, the heat noise introduced by ground spillage will plummet remarkably, hence improving the sensitivity of the telescope. The calculation results show that a moderately high noise-proof wall, which can work in coordination with the "back-illumination" model to enhance the sensitivity of FAST, can be built on the periphery of the 500-m spherical surface of FAST (Fig. 64).

(4)   Digitalization and digital signal processing

The strength of the electromagnetic wave signals, which are reflected by the reflector, received by the feeds, amplified by the low-noise preamplifier, undergo filtering and the follow-up signal amplification, and are then transmitted to the data processing terminal, can be detected and processed by electronic devices such as power meter and spectrum analysis device. We refer to the devices used for processing the amplified celestial radio radiation as the data processing terminal. The data processing terminal in the early stage generally utilized wave filters and detectors and other analog devices to measure spectrum and power, but they are limited in terms of the technical indexes they can reach and the flexibility of their usage.

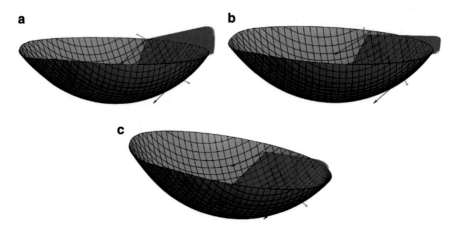

Fig. 64 The "back-illumination" model when the observing zenith angle reaches 35 degrees. *The pink part in the figure represents the main illuminated range of the feed. When the feed rotates backward to the reflector center, the heat noise it receives from the ground will decrease remarkably

Current radio astronomical terminal devices usually adopt digitalized processing. First, the analog voltage signals received are sampled and quantified to obtain a series of digital signals. As long as high and low voltages are not mixed during transmission, the complete fidelity of the signals can be ensured. Unlike the phenomenon that the signal quality degrades because of signal attenuation and noise during analog signal transmission, the fidelity of the signals can be strengthened greatly during digital signal transmission. The differences between analog signals and digital signals in transmission and storage can be seen from the following examples. For instance, the early telephones transmit analog signals, and when the transmission distance increases, signal attenuation also increases, while the signal-to-noise ratio decreases. Therefore, speaking loudly to the microphone during long-distance calls is usually required for the other side to hear clearly. The same goes for the analog televisions. In the locations far from the transmitter station, the signal-to-noise ratio of the television signals will reduce, so there will be "snowflakes" appearing in the images. By contrast, though signal loss will also occur during digital signal transmission, as long as the mixture of high and low voltages do not happen in relay and retransmission, the signals can retain their original values without losing fidelity. Therefore, exclaiming toward the microphone is no longer necessary nowadays during long-distance calls, let alone overseas calls. The receiving effects of the digital televisions are clearer as well.

There are several benefits when the radio astronomical terminal adopts the digital technology. For example, digital filters usually perform better than analog filters, and their filter function, unlike that of the analog filters, does not vary in accordance with the performance of the devices. After the digital signals are stored, all information can be retained and reprocessed through different processing approaches, and thus the signals received by the telescope can be utilized to the greatest extent.

But the analog-to-digital conversion of the analog signals will add some noises to the signals, which will decrease the signal-to-noise ratio. For analog signals of certain bandwidth, according to the Nyquist's law, if signals are sampled at a frequency corresponding to twice the bandwidth, then the data between the sampling points can be restored without loss through the data of the sampling points, which are quantified and represented by a limited number of bits. For example, for the quantification of 8 bits, there are 2^8 (256) scales in the voltage range of the analog-to-digital conversion. Such discretized quantification of voltage will introduce differences between quantified voltage and input voltage. The former can be represented by the sum of the latter and certain quantified noise. As to the input voltage with certain statistical nature, like the white noise, the average value of the quantified noise can be calculated. For the quantification of 8 bits, the quantified noise is fairly small when compared to the input signals. If there are very strong radio frequency interference signals in the observation band, then a higher number of bits in quantification are required to improve the dynamic range of the quantifying system. With regard to FAST, there are no strong radio interference signals in the observation band because a radio quiet zone is established, so 8-bit quantification is enough. When the strength of the radio sources fluctuates sharply during observation, or when there are short-time strong interference signals, maybe a higher number of bits in quantification will be needed.

During radio astronomical observation, it is usually necessary to measure the average power of the received signals within certain time intervals. The average power of the input signals is proportional to the integral average of the square of voltage. In the digital terminal, the input voltage turns to a series of discrete digital voltage signals after analog-to-digital conversion. The average power of such discrete voltage sequence is often represented by the average of the square of a certain amount of discrete voltage. Mathematically it can be demonstrated that for given sampling intervals, as long as the sample number is large enough (in other words, the average time is long enough), the integral average of the continuous function basically equals the average of the discrete sampling value. For a given average time, using shorter sampling intervals can make these two values basically equal.

(5)   Development of digital terminal and all kinds of algorithms

The FAST data processing requires the support of corresponding algorithms. The main observation modes of FAST include pulsar search, pulsar's pulse profile observation, high-resolution spectral line observation, total flow observation, and so on. The core algorithms of these observation modes involve processing the data in time domain or frequency domain, and all of the calculation procedures are predetermined. Take polyphase filter banks for example. By selecting data of certain time duration as a digital filter, its bandpass response and out-of-band rejection are far superior to those of an analog filter. Besides, with stable performance, the digital terminal can replicate the same design in the same way and it thus enhances the reliability of the data processing terminal. Another observation mode is to record the observed voltage data, which can be used in post processing repeatedly and for interference processing with other telescopes.

(6)   Overall design of the receiver

During the design process, it is necessary to optimize the overall performance of the FAST receiver system according to different observation goals. The optimization of sensitivity is its major goal for observing point sources. The optimized design of the feed can make the aperture field of the telescope as even as possible and with the least leakage, and thus reach the highest axial gain (i.e. the sensitivity of axial observation). The FAST observation band covers the frequency range from 70 MHz to 3 GHz. Over the band of 500 MHz, the sky background noise decreases, which means that the noise of the receiver has already become a large part of the system noise of FAST. Hence, for the frequency range over 500 MHz, the method of lowering the temperature of the working environment of the receiver devices is used to reduce the noise temperature of the receiver. The project team has adopted the GM refrigerator that is usually utilized in the centimeter band radio astronomical receiver, so as to make the working temperature of the polarizer part drop to 50–70 K, and that of the low-noise amplifier to 10–20 K. The FAST project will follow closely the international development of the related technology, and carry out the development of the related feed technology (e.g. phased array feed technology). With the emergence of devices with lower noise temperature, the FAST project team will further the development of receivers with lower noise to obtain higher observation sensitivity.

The development of the FAST digital data processing terminal benefits from the software and hardware development of digital signal processing and computer technology. At present, the FAST terminal adopts partially-developed high-speed ADC (analog-to-digital converter) and FPGA printed circuit board combined with the computer cluster technology. As the technologies develop, the next generation of the FAST terminal is expected to be designed and realized through the entire use of the commercial high-speed ADC, high-speed data transmission, and computer cluster, which will further improve its reliability and shorten the research and development period.

4.6   Electromagnetic Compatibility

Reading the information on the fringe of the universe requires telescopes with large aperture and high sensitivity. Therefore, many new high technologies, such as "zero" noise quantum amplifier, large structure homology with micron-level precision, and real-time processing and transmission of net bit rate data, are all developed because of the need of radio astronomy. At the same time, with the development of electronic technology and computer technology, radio telescopes utilize a large amount of electrical and electronic equipment. Besides, the electronic equipment has increasingly wide frequency band and increasingly high sensitivity, and the cable network connecting all sorts of devices become more and more complex.

> **Knowledge Link**
> **Electromagnetic Compatibility (EMC)**
> Electromagnetic compatibility, or EMC, refers to the capability of a device or a system to operate in its electromagnetic (EM) environment in accordance with the requirements and not to cause unbearable electromagnetic interference (EMI) to any other devices in its environment. For the interference produced by the electrical and electronic equipment used by a radio telescope, if no measure is taken, the normal operation and scientific output of the telescope will be severely impacted.

Hence, the issue of electromagnetic compatibility becomes more and more important.

As the world's largest and most sensitive single-aperture radio telescope at present, FAST faces great challenges and risks in its electromagnetic compatibility. The major technical difficulties are as follows: the super-sensitive FAST can reach -320 dB·W/(m^2·Hz) within the frequency band of 70 MHz–3 GHz; the observation band involves the low frequency band which is most vulnerable to electromagnetic interference; the complicated working status involves up to thousands of electrical and electronic devices, most of which need to operate at the same time with the telescope observation; and the protection threshold of radio astronomical continuous spectrum observation is about 80 dB lower than that required by the GJB151A *Electromagnetic Emission and Sensitivity Requirements for Military Equipment and Subsystem*, while the protection threshold of spectral line observation is about 100 dB lower (Fig. 65).

Therefore, the FAST project team has carried out systematic and comprehensive design and development of high performance electromagnetic compatibility for the

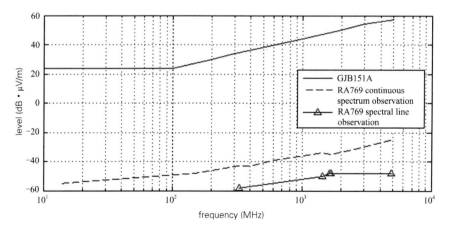

Fig. 65 Comparison between the protection thresholds in recommendation ITU-R RA.769 by the international telecommunication union and in GJB151A

super-sensitive radio telescope for the first time. By establishing the 30-km-radius electromagnetic wave quiet zone and the restricted airspace, the team has made several technological breakthroughs in the electromagnetic compatibility of FAST, which has reached the international advanced level. The realization of comprehensive electromagnetic compatibility from the air to the ground and from the radio wave environment to the telescope equipment provides indispensable support for the effective operation of FAST.

(1) High performance electromagnetic compatibility technology in areas with multi-physical fields

The project team proposed the approach of high performance electromagnetic compatibility technology in areas with multi-physical fields, and has made breakthroughs in key technologies such as partition shielding, through-wall shielding device for the shaft of large-power motors, dynamic and static shielding and combined shielding, as well as shielding structure with high shielding effectiveness for self-adaptive movable components. The shielding effectiveness exceeds the current national standard, and even surpasses 140 dB in some part of the frequency band.

The signals of celestial bodies received by the large-aperture radio telescope are extremely faint, while the electromagnetic signals produced by the equipment will easily interfere with the reception of the celestial signals by the feed, hence electromagnetic compatibility measures need to be taken to meet the rigid requirements of electromagnetic shielding in astronomical observation. Meanwhile, it is also necessary to deal with the shielding technology in strong interference electromagnetic field with a large range and multiple movable components. Such technology has no precedent in the national electromagnetic shielding field. Therefore, the following technical difficulties need to be overcome.

① Reaching electromagnetic shielding of over 120 dB

The cable-driven mechanism is one of the three self-innovations of FAST. Unlike ordinary shielded enclosures, the FAST cable-driven room has large-power servo drivers, motors and other sources of strong interference, which produce interference signals of multiple types, with great strength and wide frequency span. Moreover, the movement of the steel cables at the rope outlets of the room makes it impossible to close the wall surfaces. Therefore, the electromagnetic shielding faces great difficulty.

As the core equipment of FAST, the feed cabin is filled with a large number of electronic devices inside, including the AB axes motor, the Stewart platform six-foot motor, the controller, the processor, the feed, and the like, and a cryogenically cooled receiver is installed in the lower platform. Therefore, the feed cabin is extremely vulnerable to interference. The mass of the entire equipment is strictly limited, and the electromagnetic shielding is required to exceed 120 dB, the D level specified by China's military standards. All in all, the task is extremely challenging.

② Partition shielding technology of mechanical transmission system

There are several mechanical transmission and driving components in the cable-driven equipment, which belong to different areas and produce interference signals with different characteristics. The mechanical transmission part needs shielding and isolation, but the mechanical connection between the motor and the transmission components as well as its dynamic nature means that the conventional methods of electromagnetic shielding are not applicable. Furthermore, there is no precedent to follow in terms of high performance shielding of the mechanical transmission system.

③ Shielding structure with high shielding effectiveness for self-adaptive movable components

The complex movement and the bad environment of the movable components of the feed cabin pose a difficulty to the design, research and development of shielding structure with high shielding effectiveness for self-adaptive movable components.

With regard to the equipment interference parameters and the shielding requirements of installation locations, the project team proposed the shielding technology for multi-physical areas, including partition shielding with different indicators, through-wall electromagnetic shielding for shaft, dynamic and static shielding and combined shielding, and the like.

① Partition shielding with different indicators for multi-physical areas

Partition shielding with different indicators for multi-physical areas is applied to interference physical fields such as the large-power servo driver, the high frequency controller, the electric motor, the camera, the code disk, and the exchanger in the cable-driven shielded enclosure, as well as to the Stewart platform, the motor, the measuring equipment and the receiver inside the feed cabin.

The 120 dB shielding indicator is adopted for the electrical room which produces strong signal interference to the cable-driven system, and 50 dB is used for the motor room. Multiple levels of shielding ensure high shielding effectiveness. In the electromagnetic compatibility design of the feed cabin, the space within the feed cabin is partitioned into 3 areas for the installation of strong interference sources. For example, the shielding compartments 1 and 2 are for the driving and control equipment with the shielding indicator of over 120 dB, while the shielding compartment 3 is for the motor and driving devices with the shielding indicator of 80 dB. Moreover, the shielding of the compartment 3 is required to be realized through double layer shielding, and to surpass the national standard of 120 dB shielding effectiveness (Fig. 66).

This technology realizes highly efficient shielding of compound interference signals, successfully solves the interference problem of the large composite electromagnetic field in the open shielded enclosure, makes technological breakthroughs

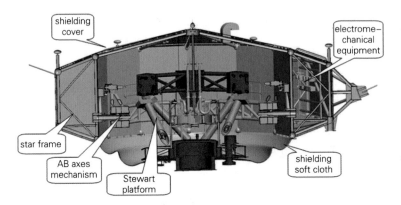

Fig. 66 Partition shielding for multi-physical areas in the feed cabin

in electrical cabin with complex movement and high shielding effectiveness, and lays foundations for the precise analysis of the signals observed by the large radio telescope.

② Dynamic and static shielding technology

Based on the dynamic and static shielding technology, a new method of dynamic through-wall electromagnetic shielding was proposed for the cable-driven shielded enclosure, and a high performance through-wall device for shaft which can prevent electromagnetic leakage was developed. As a result, the difficulty of through-wall electromagnetic shielding for the motor shaft has been solved, which ensures the effective reduction of interference of the cable-driven equipment.

For the feed cabin, scientists innovatively put forward the double layer shielding cabin design with a motion platform. The 0.8 mm thick stainless steel plates are welded to serve as the cabin cover and the partitions of the feed cabin, and the double layer shielding cloth, the rainproof layer and the supporting shelf are applied to the motion platform. After test, the feed cabin finally realizes over 120 dB shielding effectiveness, and even over 140 dB for some part of the frequency band (Figs. 67 and 68).

(2) Electromagnetic compatibility complex system with mechanical, electrical and hydraulic integration

The project team proposed the technical method of electromagnetic compatibility complex system with mechanical, electrical and hydraulic integration, and developed 2,225 large load actuators with mechanical, electrical and hydraulic integration and with high electromagnetic shielding effectiveness for the radio telescope. By innovating the structural design to reduce the replacement of components and to realize electromagnetic shielding measures for the multi-interface actuators, they ensure the automatic deformation performance requirements of the FAST active reflector.

Fig. 67 Photo of the shielding compartments of the feed cabin

Fig. 68 Shielding measures of the lower platform of the feed cabin

The active reflector of the telescope is one of the three innovations of FAST. In order to realize the formation of a 300-m-aperture instantaneous paraboloid during observation, 2225 sets of actuators are installed under the FAST reflector nodes to control their deformation. The actuators must operate at the same time with the telescope observation, and work continuously and in real time within the frequency band of below 3 GHz. The major interference sources within the actuators include the DC power supply, the motor, the motor driver, the controller, the electromagnetic valve, the oil temperature and oil pressure sensor, and the location sensor, etc. The main technical difficulties faced by the actuators in terms of electromagnetic compatibility are as follows.

①   The electromagnetic compatibility technology of large load hydraulic actuators

The loads of the reflector actuators for conventional international telescopes are usually less than 1 ton and have relatively low power. But the load of the active reflector actuators in this project is up to 7 tons, with some part up to 10 tons or even 15 tons. For such a large load, if the mechanical actuators that are frequently used at home and abroad are adopted, there will be difficulties such as short working life and large electrical power, meaning that the electromagnetic compatibility measures will have extremely large difficulties and lack feasibility. The project adopts mechanical, electrical and hydraulic integration actuators to ensure the feasibility of the electromagnetic compatibility measures of the large load actuators.

②   High performance electromagnetic compatibility technology of the multi-interface complex mechanical electrical shielded enclosure

The shielding effectiveness requirement of the FAST actuators is 80 dB, while in other countries, for example, for the 100-m Green Bank Telescope (GBT) in the U.S. National Radio Astronomy Observatory, which is the largest fully steerable radio telescope, the shielding effectiveness of its actuators is only about 40 dB.

Since each actuator is a mechanical, electrical and hydraulic integration device, the mechanical, electrical, and hydraulic interfaces, which have telescopic components, are complicated, and there are many electromagnetic interference leakage links. The small volume of the electrical cabin of the actuator not only limits the implementation of several electromagnetic compatibility measures (e.g. shielding reeds), but brings difficulties to the test of shielding effectiveness as well.

With regard to the main technical difficulties faced by the FAST actuators in terms of electromagnetic compatibility, the project team has worked hard to overcome them and innovatively proposed solutions.

①   The electromagnetic compatibility technology of large load mechanical, electrical and hydraulic integration actuators

In the system framework, the use of hydraulic actuators can realize a large load of up to 15 tons, which avoids the difficulty that the life of the mechanical actuators is reduced due to mechanical wear. After operating for a long time, the main components do not need replacement; only pumps, valves, seal rings and other small low-cost components need replacing to prolong the service life, laying good foundations for the operation and maintenance of FAST for 30 years or even longer time.

The development of mechanical, electrical and hydraulic integration actuators can not only realize a large load, but realize the shielding effectiveness of 80 dB as well. Their interfaces with digital functions such as optical fiber communication can check the status of the actuators remotely at any time and control the operation in a real-time manner, which follows closely the future development trend.

② High performance electromagnetic compatibility technology of the multi-interface complex mechanical electrical shielded enclosure

The flexible arrangement of hydraulic pipelines effectively realizes the isolation of electrical components, deals properly with the possible electromagnetic leakage from the gap caused by the movement of the telescopic rods, brings great convenience to the handling of electromagnetic interference of the electrical and electronic devices of the actuators, and greatly reduces the difficulties and costs of electromagnetic compatibility treatment. This structure not only can realize good electromagnetic shielding, but realize good environment protection as well, so it is quite suitable for bad environments, and can also suit the working environment which has high requirements for electromagnetic shielding (Figs. 69 and 70).

Scientists have successfully solved the difficulties in electromagnetic compatibility of the mechanical, electrical and hydraulic integration actuators of the FAST reflector. The related technology has also been successfully applied to the FAST actuators. The test observation of the telescope shows that there is no detection of

Fig. 69 Lab test of the actuator

Fig. 70 Field test of the actuator

interference from the actuators, which fully proves the reliability of this technical achievement.

(3)   Autonomous and continuous electromagnetic shielding test of ultra-high dynamic range and wide frequency band

The project team proposed the method of autonomous and continuous electromagnetic shielding test of ultra-high dynamic range and wide frequency band. By optimizing the software and hardware, the autonomous and continuous test capability can be realized in the range from 70 MHz to 3 GHz. The test range of the highest shielding effectiveness of the system is up to 140 dB, which remarkably enhances test efficiency and meets the strict requirements of the FAST electromagnetic shielding test.

Since the FAST requirement of electromagnetic interference suppression far exceeds the current national standard, the current national standard of electromagnetic compatibility test can hardly meet the FAST requirement of electromagnetic compatibility test. FAST embraces more than 2000 electromagnetic shielding devices, shielded rooms, and shielded cabinets, so that it is necessary to conduct rigid electromagnetic shielding effectiveness test in the whole working frequency band of FAST. The conventional measurement method mainly features a single frequency point and manual test, so completing the test of over 2000 shielding instruments and increasing test frequency points will cause a great amount of work. Besides, such test cannot show comprehensively and accurately the electromagnetic interference within the frequency band.

Compared to the general test standard, the FAST electromagnetic shielding test has special requirements which also constitute its main technical difficulties.

① Requirement of a large dynamic range. The highest dynamic range is required to be over 130 dB (while the highest shielding effectiveness required by China's military standards is 120 dB).

② Focused measurement frequency band. The measurement frequency range between 70 MHz and 3 GHz is required not to include the measurement of shielding effectiveness of magnetic field below 30 MHz, but to include the measurement of shielding effectiveness of the resonance frequency band.

③ Dense measurement frequency points. It is required that the frequency with the lowest shielding effectiveness should not be missed out, and that the shielding effectiveness—frequency curve that is nearly continuous should be drawn (according to the national standard, only 4 frequency points within the FAST observation frequency band are measured).

As to the aforementioned technical difficulties, the project team's technicians have overcome all kinds of obstacles to find breakthroughs, and developed the continuous electromagnetic shielding test system of ultra-high dynamic range (140 dB) and wide frequency band.

For the special needs of radio astronomy, especially of the FAST electromagnetic shielding test, the continuous test system of wide frequency band has been innovatively developed. Through optimization of software and hardware system, autonomous sweep frequency test can be conducted to the electromagnetic shielding effectiveness of the shielded enclosures within the wide frequency range from 70 MHz to 3 GHz. The number of sweep frequency points can be set as required; in general, 50 or 100 frequency points are tested, which surpasses the range under the current standard. The system can test shielding effectiveness as high as 140 dB (exceeding the national standard of 120 dB), remarkably enhancing the work efficiency and test quality of shielding effectiveness test and meeting the rigid requirement of the FAST electromagnetic shielding test. This also fills the domestic gap of the electromagnetic shielding test system of complex system with high sensitivity.

In conclusion, the design and the related technologies and achievements of electromagnetic compatibility of ultra-high sensitivity complex system are successfully applied to FAST. The electromagnetic compatibility measures are very fruitful, and the electromagnetic interference of the telescope per se has been effectively reduced. The FAST discovery of new pulsars smashed the past blank record of Chinese radio telescopes in the discovery of pulsars, and the development and realization of electromagnetic compatibility plays an important role in ensuring this breakthrough (Fig. 71).

Fig. 71 Field test of the
FAST test system

5 Excellent Abilities of "China's Sky Eye"

The construction of the FAST project officially started in 2011. Since then, under the full support of several national departments and of Guizhou Province, Chinese Academy of Sciences has cooperated closely with scientific research institutes, universities and corporates at home and abroad. Relying on independent innovation and taking advantage of Guizhou's unique and superior natural conditions, scientists and engineering technicians have overcome plenty of engineering difficulties, researched and developed a series of crucial core technologies, and made many technological breakthroughs in terms of program design, component development, system integration, and project implementation, and so forth.

5.1 Cable Net Structure: Active Deformation Working Mode with Ultra-large Span, Ultra-high Precision, and Ultra-high Fatigue Resistance

The FAST cable net has the world's largest span and highest precision at present, and is also the first cable net system which adopts deformation working method in the world with no precedents across the globe. Led by the needs of the FAST project, the project implementation departments have realized the matching system for the production of the high precision cable structure, which has been applied to the projects such as the stay cables of the Hong Kong-Zhuhai-Macao Bridge (HKZM Bridge), the Piliqing River Bridge of Yili, Xinjiang, and the Indian cable-stayed bridge STAR BAZZR, and has greatly enhanced China's steel cable structure manufacturing level. The relevant achievements have gained many prizes such as the "Grand Prize in the Science and Technology Awards of the China Steel Construction Society" in 2015, the "First Prize in the Science and Technology Awards of Beijing Municipality" in 2016, and the "First Prize in the Technological Innovation Awards of the Science and Technology Awards of Guangxi" in 2016.

5.2 Movable Optical Cable: High Strength, Long-Term Bending Fatigue Resistance, and Low Loss

As the first radio telescope to adopt the light-weight feed platform with flexible cable supporting structure, FAST has broken through the simple rigid supporting mode in traditional radio telescopes in which the feed and the reflector are relatively fixed. The FAST 48-core movable optical cables which have been successfully developed relying on the FAST project can meet the requirements of signal and data transmission in the flexible supporting system under the large span movement status, and also overcome the signal transmission "lifeline" difficulties in the plan of linking cable into the feed cabin. This scientific research achievement can be extensively applied to military and civil projects. The related achievements have gained many prizes such as the "Second Prize in the Awards for Scientific and Technologic Progress of Guizhou Province" in 2017, the "First Prize in the Science and Technology Awards of China Machinery Industry" in 2017, and the "Golden Prize of Good Design in China Innovation Design Conference" in 2017.

5.3 Dynamic and Stereo Measurement System: Large Scale, Multiple Targets, and High Precision

The FAST active reflector deformation measurement requires to realize real-time dynamic measurement with 2 mm positioning precision to the 2225 nodes in the

500-m scale in the wild field, and such measurement will directly influence the observation performance of the telescope. Therefore, the program of dynamic and three-dimensional measurement system with large scale, multiple targets and high precision was proposed. The large-scale and high precision real-time measurement system fills the gap of large-scale, multiple targets and high precision dynamic measurement and online metrological verification in the field of precise engineering surveying.

5.4 Electromagnetic Compatibility Measures of the Super-Sensitive Complex Telescope System

FAST is the most sensitive single-aperture radio telescope in its observation frequency band, so it is extremely vulnerable to the electromagnetic interference from its own electrical and electronic devices. Therefore, the design and measures of electromagnetic compatibility are indispensable. To reduce the interference from the telescope's own devices, the comprehensive design of electromagnetic compatibility has been finished and implemented to the telescope and its ancillary facilities. Key devices such as the 2225 actuators have all been developed and built based on the electromagnetic compatibility design, while for complex systems like the feed cabin, measures such as double layer shielding, combination of rigid and flexible mechanisms, and separate treatment have been taken. As a result, the actual shielding effectiveness surpasses the highest national standard of 120 dB. The electromagnetic shielding measures adopted can be applied to aeronautics, astronautics, ship and other systems with high requirement of electromagnetic compatibility, and have remarkably enhanced China's technological competence in the related fields.

Long March of Dream Pursuers

Haiyan Zhang, Lei Qian, Caihong Sun, Chengmin Zhang, Wenjing Cai, Aiying Zhou, Chengjin Jin, Li Xiao, Dongjun Yu, Qing Zhao, Boqin Zhu, Wenbai Zhu, Lichun Zhu, Ming Zhu, Liqiang Song, Mingchang Wu, Baoqing Zhao, Ming Zhu, Gaofeng Pan, Hui Li, Rui Yao, Youling Yue, Bo Zhang, Rurong Chen, Boyang Liu, Li Yang, Na Liu, Jiatong Xie, Yan Zhu, Hongfei Liu, Zhisheng Gao, and Xiaobing Chen

With the receiving area of approximately 30 soccer fields, the 500-m aperture FAST will maintain its status as the world-class equipment in the next 10–20 years. As it has been completed and put into operation, it will provide opportunities for astronomical development.

1 World Record

The Arecibo Observatory, located in the karst landform of Puerto Rico, had reigned supreme over half a century as the single-aperture radio telescope with the largest receiving area in the world since its construction was completed in 1963. The Arecibo telescope dish measured 305 m in diameter initially, and expanded to 350 m in 1970s.

FAST took the place of the Arecibo telescope and became the world's largest radio telescope in September, 2016. With the 500-m aperture and the receiving area equivalent to the size of 30 soccer fields, FAST has not only set a new world record in the size of single-aperture radio telescopes, but also reached a new peak in terms of sensitivity and comprehensive performance.

FAST is 10 times more sensitive than the 100-m telescope near Bonn, Germany, once known as "the largest machine on the surface of the Earth". If celestial bodies are evenly distributed across the universe, the number of targets that can be observed by FAST will increase by about 30 times. Besides, compared with the Arecibo Observatory, which was rated as No. 1 of the Top Ten Projects of the 20th Century, FAST is 2.25 times more sensitive. FAST is expected to maintain its world-class level in the next 10–20 years, attract top talents at home and abroad to undertake related leading-edge scientific research subjects, and become the academic exchange center for international astronomy. Located in the natural depression of Guizhou Province, FAST will also serve as a beautiful scientific landscape which can boost the economic prosperity and social progress of western China.

© Zhejiang Education Publishing House 2021
R. Nan (ed.), *The Sky Eye*, China's Big Science Facilities,
https://doi.org/10.1007/978-981-16-3824-4_4

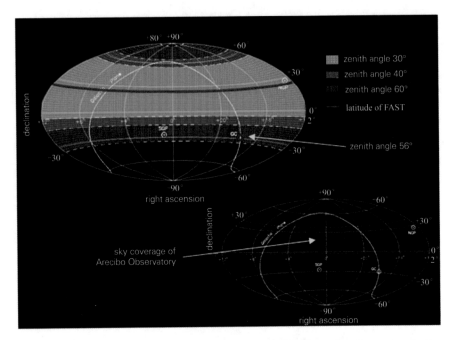

Fig. 1 Comparison of sky coverage between FAST (top left) and Arecibo Observatory (bottom right). The zenith angle of FAST during its work is 40°, twice as large as that of the U.S. Arecibo telescope

The brand new design idea and the special advantage in site location have enabled FAST to break the 100-m limit of radio telescopes and initiated a new mode of building giant radio telescopes.

The FAST project has gained extensive attention both at home and abroad. The American magazines, *Science* and *Nature*, have reported the FAST project for many times. FAST was also listed in the 2016 major scientific events appraised by *Nature*. In December 2016, voted and appraised by the CAS (Chinese Academy of Sciences) and CAE (Chinese Academy of Engineering) academicians, the news of "World's Largest Single Aperture Radio Telescope Completed and Put into Operation in Guizhou" was ranked in the top ten news reports of national scientific and technological progress in 2016 in China (Fig. 1).

2 Discovery of New Pulsars

Since the construction of FAST was completed on September 25th, 2016, the National Astronomical Observatories of the CAS (NAOC) has led several national institutes and worked in close cooperation with the FAST scientific and engineering team to enable FAST to realize several observation modes including drifting, tracking,

motion scanning, and knitting scanning, etc. The progress of commissioning exceeds expectation and the international practices of large equipment of the same kind, and FAST has begun to generate systematic and scientific outputs.

On October 10th, 2017, CAS hosted and released the first set of achievements made by FAST, including the discovery of 6 new pulsars. The first among them is numbered J1859-01 (also known as FP1-FAST pulsar #1). Its rotation period is 1.83 s, and it is calculated to be 16,000 light-years away from the Earth. It was discovered on the southern sky of galactic plane through drifting scanning on August 22nd, 2017, and verified by the Australian Parkes Telescope on September 12th, 2017. This was the first-ever discovery of new pulsars by China's radio telescope, which was reported and followed by several authority media including CCTV News, *People's Daily*, *Science and Technology Daily*, *China Science Daily*, *Economic Daily*, *Guangming Daily*, *China Youth Daily*, *Jiefang Daily*, and *Wenhui Daily* as well as xinhuanet.com, people.cn, chinanews.com, and chinadaily.com.cn. "FAST first discovers pulsar" was chosen as one of the CAS annual scientific and technological innovation spotlight achievements in 2017. "FAST discovers several pulsars" was selected into the top ten national scientific and technological news reports in 2017 (Fig. 2).

The search and discovery of pulsars is one of the core scientific goals of FAST. By the end of December 2018, FAST has observed 70 pulsar candidates, among which 53 have been verified as newly discovered pulsars. China has reached the world-leading level in the field of discovering pulsars. In the newly discovered pulsars, one is the faintest radio millisecond pulsar in the high energy pulsars discovered so far. Other large facilities in the world have not searched its radio pulse previous to its discovery by FAST, which fully demonstrates the advantage of FAST in sensitivity. Such discoveries have broken the past blank record of China's telescope in discovering pulsars.

Fig. 2 Effect picture of FAST detecting pulsars

3   Future Prospects

Astronomy is a frontier science which breeds major and significant innovative discoveries, and also a strategic high ground which promotes scientific and technological progress and innovation. The construction and operation of FAST are of great importance for China in making major innovative breakthroughs and accelerating innovation driven development at the leading edge of science.

There are a large number of pulsars in the Galaxy, but due to their weak signals which are easy to be flooded by artificial electromagnetic waves, now only a small part of them have been observed. The super-sensitive FAST is the ideal equipment to discover pulsars. Its discovery of pulsars in the initial stage of commissioning owes to the fruitful early scientific planning and personnel and technology reserves, demonstrates preliminarily the independent scientific innovation ability of FAST, and ushers in an era in which China's large scientific facility system of radio band makes original discoveries intensively. In the future, FAST is expected to discover more millisecond pulsars with time accuracy, and make its original contributions to the detection of gravitational waves through pulsar timing array (Fig. 3).

The completed FAST is going to play a huge and indispensable role in solar-terrestrial environment research, search for extraterrestrial intelligence, national defense construction, national security and other aspects with significant national needs. Its construction and operation will also benefit the economic prosperity and social progress of western China, which is in line with China's overall strategy of regional development. The emergence of FAST turns the remote karst mountainous area in Qiannan, Guizhou, to an international astronomical academic center that has captured the world's attention, which offers a new window for presenting Guizhou

Fig. 3 Pulsar

Province to the world. The construction of the astronomical scientific education base centered on FAST will promote China's scientific education, develop teenagers' scientific innovation abilities, enhance the publicity to the public and the policy makers, and serve the long-term strategic goal of boosting the country through scientific and educational advances.

Major Events in FAST History

Jun. 1994	The Beijing Astronomical Observatory (i.e., the National Astronomical Observatories presently) set up the large radio telescope (LT) research group, began site selection, and started its 13-year-long cooperative preliminary research
Mar. 1998	The complete concept of FAST was put forward
Apr. 1998	20 national scientific and research institutes established the FAST project committee
Mar. 1999	The "Large Radio Telescope FAST Pre-Research" started as one of the first series of major programs of Knowledge Innovation Project
Jan. 2005	The "Overall Design and Key Technology Research of the Giant Radio Astronomical Telescope (FAST)" started as the Transdisciplinary Key Program of the National Natural Science Foundation of China (NSFC)
Sep. 2005	The project was successfully accredited in the national large research infrastructures "FAST Proposal Panel Review Conference" organized by CAS
Mar. 2006	The CAS Bureau of Basic Science hosted "International Review and Advisory Conference on FAST", and suggested that the project should apply for its establishment and start construction as soon as possible
Jul. 2007	The National Development and Reform Commission (NDRC) approved the official establishment of the FAST project
Oct. 2008	NDRC approved the FAST feasibility research report
Dec. 2008	The FAST project laid the foundations
Feb. 2009	CAS and the Guizhou Provincial People's Government approved the initial design and estimate budget of the FAST project
Mar. 2011	The construction of the FAST project officially commenced
Jan. 2012	The 973 Program of "Frontier Astrophysical Topics in the Radio Band and A Proposal for the Early Scientific Studies of FAST Project" started
Dec. 2012	The excavation began in the site of FAST, and the slope treatment engineering received acceptance check
Oct. 2013	*Measures for the Protection of the Radio Quiet Zone for the Five-hundred-meter Aperture Spherical Radio Telescope in Guizhou Province* began to be implemented
Dec. 2013	The steel structure of the ring beam of FAST was completed successfully

(continued)

© Zhejiang Education Publishing House 2021
R. Nan (ed.), *The Sky Eye*, China's Big Science Facilities,
https://doi.org/10.1007/978-981-16-3824-4

(continued)

Nov. 2014	The construction and installation engineering of the FAST feed support towers were completed and checked
Feb. 2015	The FAST cable net engineering was completed
Nov. 2015	The first lifting of the acting FAST feed cabin was successfully finished, and its parking platform passed the acceptance check
Mar. 2016	CAS and the Guizhou Provincial People's Government jointly approved the adjustment to the initial design and estimate budget of the FAST project
Jun. 2016	The FAST comprehensive wiring engineering passed the acceptance check, and the main body of the feed cabin (focus cabin) was completed
Jul. 2016	The hoisting and installation of the FAST reflector units were finished, and the main body of the FAST project was completed
Sep. 2016	The ultra-wide band receiver was successfully installed, and FAST finished its first pulsar observation
Sep. 2016	The FAST project was completed and put into operation
Aug. 2017	FAST discovered new pulsars, which made a breakthrough by smashing the past blank record in the discovery of pulsars by China's radio telescopes
Feb. 2018	FAST discovered millisecond pulsars for the first time

Printed in the United States
by Baker & Taylor Publisher Services